D0349221

Fundamentals of Electricity

JOEL GOLDBERG, Ph.D.

Macombe County Community College

PRENTICE-HALL, INC., *Englewood Cliffs, New Jersey 07632*

TK
145
.G55
1981

Library of Congress Cataloging in Publication Data

GOLDBERG, JOEL, 1931-
    Fundamentals of electricity 85025

    Includes index.
    1. Electric engineering.   I. Title.
TK145.G55   621.31   80-20000
ISBN   0-13-337006-2

*Editorial production supervision
and interior design: Karen Winkler
Manufacturing buyer: Joyce Levatino
Cover design: Jerry Pfeifer*

© 1981 by Prentice-Hall, Inc., Englewood Cliffs, N. J. 07632

All rights reserved. No part of this book
may be reproduced in any form or
by any means without permission in writing
from the publisher.

Printed in the United States of America

10   9   8   7   6   5   4   3   2   1

Prentice-Hall International, Inc., *London*
Prentice-Hall of Australia Pty. Limited, *Sydney*
Prentice-Hall of Canada, Ltd., *Toronto*
Prentice-Hall of India Private Limited, *New Delhi*
Prentice-Hall of Japan, Inc., *Tokyo*
Prentice-Hall of Southeast Asia Pte. Ltd., *Singapore*
Whitehall Books Limited, *Wellington, New Zealand*

Contents

4
Circuit Laws   **50**

5
Series Circuits   **70**

6
Parallel Circuits   **85**

7
Combination Series-Parallel Circuits 98

8
Magnetism 118

9
Electromagnetic Inductance 134

10
Generation of Voltage 154

11
Capacitance   176

12
Reactance and Impedance   189

13
Conductors and Transformers   211

18
Direct-Current Power Sources   **312**

19
Schematic Diagrams   **325**

20
Control of Electric Power   **338**

Index   **355**

Preface

One of the most significant discoveries in the history of the industrial world was the machine. People used the machine in order to save energy and/or to be more productive. Added to the application of the machine was the discovery of electricity. Electricity had the capability of providing the power with which to operate the various machines, and so it was put to work. The first chapter of this book reviews some of the history leading up to the applications of electrical energy. Present day applications are presented as well as the effects of these applications on our lives.

The next chapters present electrical theories; basic applications are also discussed. This will lead to a fundamental understanding of the principles and how the world in which we live is able to use electric power. The final section of the book presents material covering application for control of electric power. The devices used to illustrate these control applications are familiar to all of us. They are found in the automobile, refrigerator, air conditioner, heating, and lighting systems. These applications are also found in equipment used in industry.

This book is designed to aid those who require a basic understanding of electrical concepts and applications. It is also offered as a text to those persons who require knowledge of electrical theory but may not wish to pursue the study of electrical theories as a vocation. The writing and vocabulary are at a high school, first year technical school, and adult education level.

The material presented in this book will provide the reader with a fundamental understanding of electrical power applications. This knowledge may be applied to almost any type of electrical circuit, and will provide the foundations of knowledge required in order that one may be functional in our highly technical industrial society.

The author wishes to acknowledge the assistance of others in the preparation of this book. Several manufacturers of electrical equipment have provided material and illustrations which have helped make this a better book. The support, encouragement and assistance of my wife, Alice, has done more than one realizes in the preparation of this material. To all of these people I say "Thanks."

Joel Goldberg

A Bit of History

From the beginnings of the world as we know it, people have needed to work. Work is a necessity for survival. As a result, people made a variety of discoveries and developed new ideas. Most of these ideas are taken for granted today. Early scientists and others studied the world in which we live; the important things they learned became scientific laws and rules. Many of these were refined, and often new scientific laws were identified as the studies continued over many years. Many of these laws were then debated and discussed by other people working in related fields. From these discussions came a series of scientific laws and guidelines that we use today.

The period of years between 1775 and 1825 was one of intensive scientific discovery. It was during this time span that Ampere, Ohm, Volta, and Oersted identified rules about electrical phenomena. These rules and laws are still used.

Early scientists did not really know what to do with electricity. It took many years of experimentation before electricity was put to work. During the period of time that these discoveries were being made, the industrial revolution was also going on. People have always wanted to make life easier for themselves. Early production was very slow. Most items were custom made, one at a time. Parts were not interchangeable. The development of the wheel early in the history of our world increased the workers' load capability. Now items could be carried on a cart or wagon. More goods could be moved with less human effort. Efficiency improved because less effort was used to do a specific amount of work. The wheel was adapted to work with machinery. Soon water power was utilized to turn a power shaft. Belts were attached to the power shaft and used to turn machinery. This, of course, only worked in areas where there was a supply of running water.

In the British Isles the steam engine was developed. Dependence on running water was reduced because the steam engine had a storage tank for its water. Production was able to move away from rivers and streams and into almost any other part of the world. Steam also did the work formerly done by animals, as the steam engine was used in the transportation field. Steam, however, had its limitations. Engines were large and required constant attention. Boilers used to produce the required steam required attention as well. There were alternative ways available. Adaptations of these have developed the modern industrial world we know today.

People have always wanted to perform some form of work. Many of us, being naturally lazy, want to perform this work with a minimum of effort. This attitude has led to many inventions that have tended to make our lives easier, thus accomplishing our objectives. The development of electrical laws and the adaptation of these laws have enabled us to increase our capacity to do work. We are able to do this work with less and less personal effort.

PUTTING ELECTRICAL POWER TO WORK

Work is related to the application of electrical energy. We call this use electric power. The creation of electric power was done by people early in the industrial revolution. The application of electric power has increased to the point where our society is almost totally dependent upon electric power. Take a moment and look around you at home and at work. How many of the devices you are using for work or comfort are powered by electrical energy? How many use other forms of energy that are not dependent on electricity? If you are typical of the rest of us, almost all the various devices you have named use electric power in one form or another. The form may appear to be indirect, such as that identified by a wall plug outlet. It may also be more direct, like that found in a battery-operated watch or radio. As you probably have noted, our lives are very dependent upon electrical power.

This book will present the concepts of electric power and show applications of this power. You will discover that electric power used in the home, the automobile, or in industry has the same roots. The terms and rules are the same regardless of the application. You will find that once you recognize the rules and basic components used in electrically powered devices you will have a basic understanding of how each device works. You should be able to apply the rules in any situation and come away with a reasonable understanding of how the machine or device works, that is, how electric energy is utilized to make the device do some form of work.

Electric power does not appear as if by magic. It must be created by people. Once this electric power is created it has to do some work. The creation of this form of power is unusual in that it cannot be collected and stored for periods of time. It has to be used almost immediately as it is produced. The exceptions to these statements are found in devices such as batteries. Here electrical energy is stored until it is needed. Batteries, however, have their limitations. The greatest limiting factor relates to their physical properties of size and weight. Therefore, batteries at this time have limited application for electric power production.

Recent statistics from U.S. government sources show that world production of electric power was greater than 5,185,000,000,000,000 watts for one single year! The watt is the unit used to measure electric power. This huge quantity should offer some idea as to the extent to which our lives depend upon electrical power. This power was created at a large number of electric generating facilities. The electric power then was distributed over a network of wires to the consumer. Consumers include the home, government, business, and industry.

Each consumer of electric power has its own familiar devices. In the home, these would include lighting, cooling, cooking, refrigeration, entertainment, and hobbies. Heating could also be included. If the home does not have electric heat, electric power is used to operate the fans and control the heating system. Many homes are dependent upon electric power to pump water for use in the house. There are too many other applications to list all of them.

In industry, electric power operates the machines. Electric power is also used to control the operating cycles of the machinery. Therefore, electric power is used in two different ways. The largest amount of power is normally used by the electric motors. Only a small percentage is used in the control function. Other industrial applications include smelting, melting, fusing, welding, and plating processes. Here, as in the home, there are literally thousands of specific applications.

Electric power has replaced wind, water, and steam power as the principal form of energy with which we produce work. In many instances, electric power also replaces mechanisms found on machines.

An excellent example is the adding machine-calculator. Early versions of these machines used gears and levers to provide numerical answers. They were relatively slow and could become very large. The development and refinement of electrical principles over the years between 1915 and 1970 gave us the vacuum tube, the transistor, and the integrated circuit. Today's adding machine-calculator is a very personal device. Some of these are so small that they take up the space of a business card. These electronic devices are capable of storing information as well as processing information. Their cost is a fraction of the cost of an earlier mechanical device. Time for operation is also reduced to a minimum owing to the faster speed of operation. Consumption of electric power is also reduced to a very low amount when compared to either a mechanical device or to some of the early electronic computers.

SUMMARY

The discovery and application of electric power was not an overnight development. The history of electicity goes back to the Greek civilizations. Electrical mysteries were discovered over 2,500 years ago. It was not until the late 1700s that the scientists of the day found ways to utilize this phenomena. Some of these poeple, Ampere, Ohm, and Volta, were just a few, gave us rules and laws that govern electrical energy. Later, scientists like Faraday and Edison found ways to apply the rules and come up with practical applications. The applications were refined and improved until modern-day motors, generators, and lights were developed. In addition, we now have electronic devices, such as the vacuum tube, transistor, and integrated circuit. These play a very significant role in our way of life. We are now able to use large quantities of electrical power to perform work. We also can use electric power in smaller quantities to control the larger quantities required to operate the machinery.

Electricity is a form of energy that our society depends on for survival. Its applications and control are essential for the highly technical society in which we function. An understanding of the production, distribution, and application of electrical power is important and valuable in this society. The knowledge of basic concepts is essential as utilization of electric power becomes more diverse. This knowledge will be applied to new technology in order to understand how devices and systems work. The applications keep changing. The rules, however, remain the same. Learn the basics and then relate them to the large variety of electrically operated devices we have. If success is based upon understanding, then knowledge of the rules should make us successful in our work.

2

Chemical, Solar, and Thermal Creation of Electrical Power

The study of electricity can only be successful if one has a clear understanding of terminology. The terms and phrases are used to explain how various electrical devices and circuits work. This chapter covers how electricity is produced. Some of the various ways used to create electricity are illustrated. When the means of production are understood and the terminology has meaning, those entering the field of electricity will have a better understanding of what is happening. This understanding will relate to all applications of electricity as they are encountered.

WORDS AND PHRASES

Four basic terms are used to discuss electricty. These are voltage, current, resistance, and power. All are related to each other. Each can also be used by itself to describe one aspect of electricity. Learn each of them. Later, in Chapter 3, you will be shown how they interrelate.

Electricity can be explained in several different ways. One way is to show a similar nonelectric system. The related parts are discussed to show the similarities between the nonelectrical system and an electrical system. One is not able to see electricity. It is difficult to put the terms in a form that may be visualized. This section of the book uses a fluid power system as an illustration to help put the electrical terms in a visual form.

Figure 2-1 illustrates a simple closed-loop fluid power system. At one end of the system is a pump. The pump creates a pressure in the

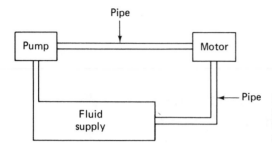

Figure 2-1. A closed-loop fluid power system. Fluid is pumped from the supply to the motor in order to perform work.

system. The pressure causes the fluid to move by creating a difference in fluid pressure in the system. The only way the fluid is able to move is through the pipes connecting the pump to the motor. The motor is shown as representing some kind of work being done. The fluid cannot collect at the motor. This would build up a back pressure and the pump could become overloaded and stop. This problem is solved by adding a pipe that returns the fluid to the pump. The fluid is reused in this type of system. The speed of the motor depends upon the movement of the fluid in the system. Now let us look at how this system is similar to an electrical system.

Voltage

In nature, matter can neither be created nor destroyed. It is possible to change the form of matter in order to do some work. This is accompliished by means of an electrical pump. This pump may be a generator, a battery, a solar cell, or any of many other devices. The electrical pump operates in a manner very similar to the fluid pump. Its purpose is to produce a difference in electrical charges. This difference in charges is measured from the outlet of the pump to the inlet of the pump. In a fluid system this difference in charges is called fluid pressure. In an electrical system it is called the *electromotive force*. Electromotive force is abbreviated EMF.

EMF is measured in *volts*. This term is shortened from the word voltage. Voltage is defined as a difference in electrical pressure, or EMF, between any two points in an electrical circuit. Common usage of the term volts has created a misunderstanding of this term. Often one says, "There are 120 volts in the circuit." What is meant is that there is a difference of electrical pressure of 120 volts between two specific points in the circuit. What is omitted is identification of a reference point when the circuit is discussed.

Voltage, when measured, is measured from one point in the circuit to another point in the same circuit. For example, a fluid pressure exists across the motor when the pump is operating. A pressure drop, or difference, exists when measured from the inlet pipe to the outlet pipe. This is illustrated in Figure 2-2. Pressure gauges are in-

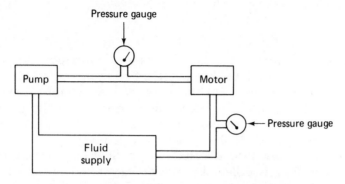

Figure 2-2.   Fluid pressure is measured by use of pressure gauges connected to pipes on each side of the motor. The pressure is the difference between the two readings.

serted in the two pipes. Each has its own reading to show pressure at that point. A pressure difference is indicated. This is the fluid pressure at the motor. Another way of saying this is that the pressure at the motor is the difference in pressure across the connections to the motor.

This same situation occurs in an electrical system. Electrical pressure in a circuit is measured as a difference in EMF between two points in the circuit. This is illustrated in Figure 2-3, which shows the

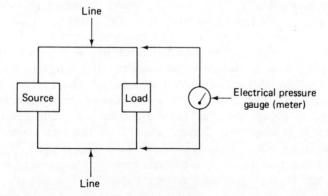

Figure 2-3.   Electrical pressure is measured by placing the leads of the meter to wires connected to the motor.

similarities between a fluid system and an electrical system. The pump and the *source* are doing the same job. The electrical circuit uses *lines*, or wires, instead of pipes to connect the source to the work. Instead of identifying the work-producing device, it is called a *load*. The electrical system requires a source of energy, a load, and lines, or wires, to connect the two devices.

Current

Electrical pressure when it stands alone is called potential electricity. In other words, it has the potential to do some work. The ability to do the work is there, but the work, or load, is not turned on or connected at the time. The electrical charges created by the source are in a state of readiness to perform some kind of work. Another name for this type of electricity is static electricity.

When the work load is connected, a flow of electrical charges occurs. This flow is called the *electric current*. It is similar to the movement of the fluid through the pipes in the fluid system. In both systems the flow is from the source (or pump) through one set of lines to the work load and then back from the work load to the pump. In the fluid system this flow can only move in one direction. In an electrical system there are two kinds of flow systems. One of these is similar to the fluid system. It is called *direct current*. In this system the electrical current flows from the source through the load and back to the source. The electron flow is always in the same direction. The term direct current is shortened to *dc* in the materials used to describe it.

Another type of source develops an electrical current that changes direction periodically. This kind of source is called an *alternating-current* source. Its name is shortened to the letters *ac*. In an ac circuit the electrical current starts, rises to a maximum value, and then drops to zero. Then the current reverses direction and repeats the same cycle, rising to a maximum value and then dropping to a no-current-flow condition. After one complete reversal, the cycle is repeated and continues in this manner. The two types of electrical currents are illustrated in Figure 2-4. Part (a) shows graphically the electrical current flow in a dc circuit. Current starts from the zero point and rises to a maximum value. It remains at that level as long as the circuit is on. Part (b) shows an ac circuit. This presents in graphic form the rise and fall and then reversal of electrical current in this circuit.

This cycle of ac is repeated a given number of times per second. The rate of this repetition is called the frequency of the electrical current. This term has been related to a standard reference of time, the second. The frequency of an electrical current is described in terms of "cycles per second." This term has become obsolete in recent years. International agreement in the scientific world has

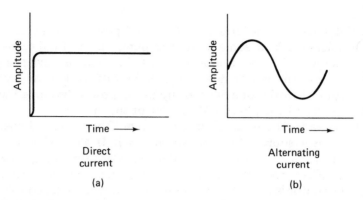

Figure 2–4. Electrical waveforms for (a) direct current and (b) alternating current.

accepted the term *hertz* to describe cycles per second. The correct way of stating frequency is illustrated by the sentence, "The frequency of the electrical current is 60 hertz."

In a fluid system the flow of the fluid may be measured. This is done by inserting a flow meter in one of the pipes. The meter measures the amount of fluid moving past this point in the circuit. Electrical systems measure the movement of the electrical current. This is done in a very similar way. An electrical flow-measuring device is called an ammeter. It measures the flow of electrical current in units of the ampere. The ampere is determined by counting the quantity of electrical charges that move past a given point in the circuit in 1 second. This number of electrical charges is very large. It is meaningless to the techncian. The *ampere*, or amp, is used to describe the electrical current movement in a circuit.

Resistance

A mechanical system for providing power has to overcome friction. Only when this is done is it possible to do some work. The same holds true in an electrical system. A different term is used to describe electrical friction. This term is *resistance*. Resistance is the opposition to the flow of current in a circuit. It is measured in units of the *ohm*.

Resistance is able to reduce the flow of an electrical current. Resistance is used to control the quantity of electrical current flowing in a circuit. Resistance may be overcome by use of a more powerful source in the circuit. There is a relationship between each of the quantities of electricity presented in this section of the book. This relationship is discussed in detail in Chapter 3.

Power

Electrical energy is described in terms of power. Electrical power does physical work of some kind. The pump in the fluid system develops energy and forces a movement of the fluid through the pipes in the system. The motor does some amount of work in this system. The amount of work done depends upon the power developed by the pump, the size of the pipes, and the size of the motor.

The same is true in an electrical system. The source of energy is called a power source. In some situations it is also called a power supply. Its purpose is to provide the necessary operating power to the load. Wires are used to carry the electrical current instead of pipes. The motor is replaced by a more universal term called the *load*. The load may be a motor, but not in all cases. Here, too, the amount of work is dependent upon the power source, the size of the wires, and the ability of the load.

The unit used to measure electrical power is the *watt*. Electrical power is the time rate for doing electrical work. This is developed by determining the amount of electrical current that is required to perform some given task and the amount of electrical voltage required to make the current move in the circuit.

A typical time frame for electrical power is the hour. Many applications of the measurement of power used to perform work use the watt-hour as a reference. This term means 1 watt of power doing work for 1 hour. This is a very small amount of work. A more common term used is the *kilowatt-hour*. Kilo- is a prefix representing 1,000. The phrase kilowatt-hour means 1,000 watts of power for 1 hour.

Terms and Notation

The scientific and industrial world uses certain terms that are abbreviations. Common usage of these terms means that it is important that persons working in the field understand them. Some of these have been introduced and explained. These are the volt, the ampere, the ohm, and the watt. A review of these terms shows that the volt is the unit of electromotive force, the ampere is the unit of electrical current, the ohm is the unit of electrical resistance, and the watt is the unit of electrical power.

In many cases these terms require additional notation. This is most often true when the terms are used to describe either very large or very small quantities. When this is required, prefixes are added to the words. Prefixes help to describe the total value of the amount in a nonconfusing manner. There are two groups of prefixes. These are both shown in Figure 2-5. Those that show larger quantities are listed

Prefix	Symbol	Value	Scientific Notation
tera	T	1 000 000 000 000	10^{12}
giga	G	1 000 000 000	10^{9}
mega	M	1 000 000	10^{6}
kilo	k	1 000	10^{3}
milli	m	0.001	10^{-3}
micro	μ	0.000 001	10^{-6}
nano	n	0.000 000 001	10^{-9}
pico	p	0.000 000 000 001	10^{-12}

Figure 2-5. Prefixes used in electrical work. Each of these represents a quantity based upon units of tens.

first. One common prefix is the term *mega*. It is the same as 1 million. Another one is the *kilo*. Kilo means 1,000. These are shown in the chart both in word form and the power-of-ten form.

One will also find several prefixes used to indicate values of quantities less than 1. Three major prefixes in this group are *milli-*, *micro-*, and *pico-*. These show decimal parts of a whole number. Milli is equal to 1/1,000, micro shows 1/1,000,000 and pico is used to show 1/1,000,000,000,000. These are also shown in powers-of-ten form on the chart.

When using these prefixes, the common accepted form is to say 1.4 kilowatts instead of 1,400 watts. This is often abbreviated to 1.4kW. The chart shown in Figure 2-6 shows letter abbreviations for

Electromotive force	Pressure	Volt	E
Electron flow	Current	Ampere	I
Opposition	Resistance	Ohm	R
Power	Work	Watt	P

Figure 2-6. Terms used to represent electrical quantities. The terms, their representative symbols, and the values are given.

the terms voltage, amperage, resistance, and power. These letters are combined with abbreviations of the prefixes discussed earlier. The phrase 1.4 kilovolts is often shortened using this format to 1.4 kV. Another example is the use of 10 mA instead of using 10 milliamps when describing current flow in a circuit.

Work Units

A resting body requires some effort to get it moving. If this body is in motion, it tends to continue moving in a straight line. A force is required to change the state of motion of the body. The force may either push or pull the body when it acts upon it. The term used to describe the quantity of this force is described in mechanical terms as the pound.

The force that acts upon a body performs work. In this case work is the product of the amount of force and the distance the force moves the body. Work is therefore described as the amount of force required to move a body some specific distance. In mechanical terms this is described in units of the foot-pound. A 5-pound weight lifted 2 feet requires 10 foot-pounds of work.

Power is the rate at which work is done. Different people will probably require different amounts of time to perform the same amount of work. The formula for mechanical power is:

$$\text{Power} = \frac{\text{work}}{\text{time}}$$

The unit of mechanical power is foot-pounds per second. Five hundred and fifty foot-pounds per second is equal to 1 horsepower.

In electrical circuits the electromotive force and the electric current flow are used to determine the amount of power. Electric power is the product of the electromotive force and the quantity of electrons that flow in a standard time frame of 1 second. The net result is a measure of electric power. This is called the watt.

Electrical energy may be converted into mechanical energy. Mechanical energy may also be converted into electrical energy. The units of mechanical power and electrical power may be equated and interchanged in the following manner:

1 horsepower = 746 watts (0.746 kW)

1 kilowatt   = 1.34 horsepower

GENERATION OF ELECTRICAL ENERGY

Electrical energy may be produced in any one of several different ways. Some of the more common ways include chemical cells, electric generators, heat, and sun or solar systems. Each of these systems are in use at the present time. The balance of this chapter is used to explain how each type of system is used to produce electrical energy. The exception to this statement is that generation of electrical energy by means of electromechanical generators is covered in the following chapter.

Chemical Cells

A common means of producing electrical energy is the use of a chemical cell. The cell is defined as a single unit capable of producing electrical energy. Cells may be, and often are, wired together to increase the amount of electrical energy available for use. When this is done, the combination of cells is called a *battery*. Wiring combinations and resulting energy levels are discussed at the end of this chapter.

Cells used to produce electrical energy are often described as either *wet* or *dry* cells. In either case they work in the same basic way, as shown in Figure 2-7. The cell contains three different units.

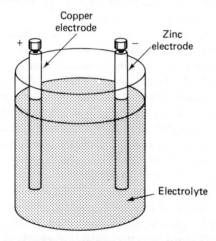

Figure 2-7. A chemical cell. It consists of two electrodes and an electrolyte material.

These units are an electrolyte material and two electrodes. The electrodes are always made of two different materials. One electrode will collect positive (+) charges from the interaction of the electrode and the electrolyte. The second electrode will collect negative (−) charges.

In the study of electricity there are two accepted methods of presenting material related to the movement of an electrical current. One method is to consider the movement of the negative charges in the system. These negative charges are called *electrons*. Electron movement in an electrical circuit is from the negative connection on the power source, through the external circuit and load, and back to the positive connection on the power source. This is illustrated in

Figure 2-8. When these electrons are moving, they form an electric current. When they are moving in a circuit, work is being done in the circuit.

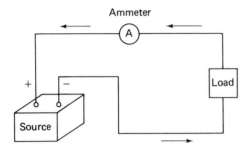

Ammeter

Figure 2-8. The flow of negatively charged electrons in the circuit. This flow is from the negative terminal of the source, through the load, and to the positive terminal.

The second method of describing the flow of electrical charges is to look at the movement of the positive charges in the circuit. The positive charges move from the positive connection on the power source, through the external circuit, and then back to the negative terminal of the power source. This charge movement produces the same results as does the flow of electrons. It seems to be a matter of personal choice as to which of the two charge paths one wishes to follow. This book uses the movement of the negative charges as a means of explaining how electrical circuits work. This is called *electron current flow*. Electron current flow is from the high concentration of negative-charged electrons found at the minus (-) terminal of the power source, out through the external circuit and load, and then to the positive terminal (+) of the power source. The positive terminal has a shortage of electrons. It readily accepts all that it is capable of handling from the negative terminal of the power source.

Dry cells. Several different kinds of dry cells are available at this time. The oldest of the group is described as a *carbon-zinc* cell. The carbon-zinc cell is illustrated in Figure 2-9. It consists of a carbon rod housed in a zinc can. The rod is surrounded by an electrolyte material. This cell is capable of producing 1.55 volts of electricity when it is fresh. Electrical current capacity of the cell depends upon the size of the two electrodes. This cell has a shelf life of about 1 year if it has not been used. It is not considered to be rechargeable.

Another common cell is the *alkaline* cell. This cell uses electrodes of manganese dioxide and zinc. It is capable of producing about twice as much electrical power as the carbon-zinc cell. This cell will handle heavier electrical currents, also. It will do this for longer periods of time than will the carbon-zinc cell. Its terminal voltage is also 1.55 volts. This cell is also not normally rechargeable.

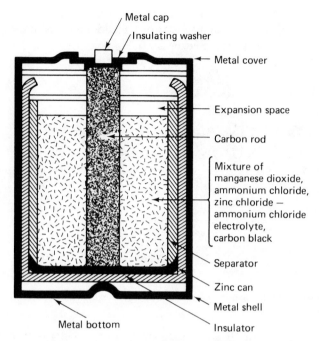

Figure 2-9. Construction of a carbon-zinc dry cell.

A third type of cell used today is the mercury cell. One of its electrodes is made of mecuric oxide and graphite. The other electrode is zinc and mercury. The terminal voltage of this cell is 1.34 volts. This cell is the most expensive of the three cells discussed so far. It has the ability to provide a steady voltage and current over its operational life. It also has the characteristic of holding its rated voltage until it is almost fully discharged. It also is not considered a rechargeable type of cell.

Another dry cell is the *nickel-cadmium* cell. Its output voltage is 1.25 volts. One of its electrodes is nickel and the other is cadmium. This cell is the most expensive of the dry cells. It has the ability to be recharged. Recharging permits almost unlimited life for the cell. It is commonly referred to as a ni-cad cell.

One additional type of cell is the gel cell. These cells are rechargeable. They are sold under a variety of trade names. One of these is the Gould Gelyte® battery. This battery and other gel cells have the advantage of being lower in cost than the ni-cad battery. It has another advantage in that its cells cannot be permanently reversed in polarity, as in the case with a ni-cad battery. The disadvantage of this type of cell is its size and weight. If one required a lightweight, longer-life battery with a high discharge rate and rapid recharging characteristic, then the ni-cad battery should be considered. The gel cell battery is illustrated in Figure 2-10.

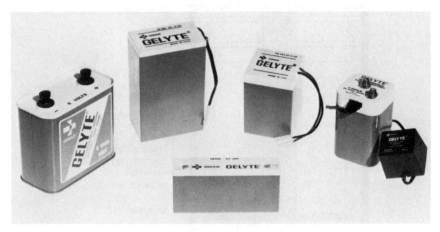

Figure 2-10.   The gel cell is available in a variety of sizes and voltages. (Courtesy Gould, Inc.)

Wet cells. A major disadvantage of most dry cells is that they are not rechargeable. Once the chemical energy is exhausted, the cell must be replaced. The discovery of the wet cell overcame this drawback. The wet cell is a storage cell. It is capable of storing electrical energy. When in use, the chemical action of the cell provides large quantities of energy. This cell may be recharged, or reenergized, by applying an external source of power to its terminals. When this is done, the chemical action of the cell is reversed. The nominal voltage available at the electrodes of the wet cell is 2.1 volts.

One of the electrodes is made of a lead peroxide mixture. The other electrode is made of lead. The electrolyte in the cell is an acid. The electrodes are shown in Figure 2-11. The use of grids instead of solid sheets makes the cell easier to produce. It also allows production of a cell with a large capacity to store energy. In a practical cell the plates are separated by an insulating material to keep them from touching each other. If the plates were to touch, the cell would have a short circuit. In this situation the voltage between the electrodes would drop to zero and the cell would be considered "dead."

Battery life is a consideration in the selection of any battery. Battery life is directly related to the amount of electron current and the length of time that the maximum current will flow from the battery to the load. Batteries are rated in units of the ampere-hour. This figure tells the user the amount of current available, usually on

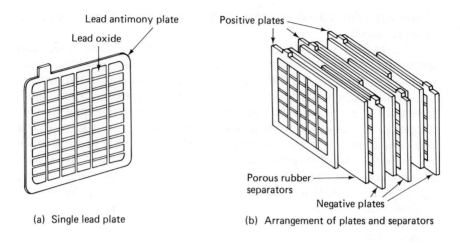

(a) Single lead plate

(b) Arrangement of plates and separators

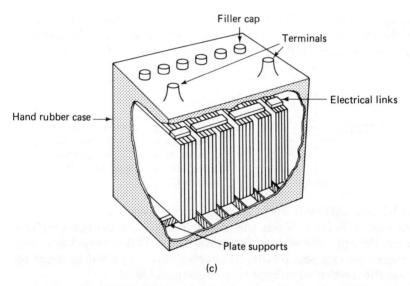

(c)

Figure 2-11. Construction of a lead-acid wet cell. Six sets of cells are used to produce a 12-volt automobile battery.

the basis of 1-hour activity. For example, a 100 ampere-hour rating will provide 100 amperes of current for a period of 1 hour. This same battery will provide 10 amperes of current for 10 hours, or any other combination that will produce the 100 ampere-hour figure.

A second factor to consider regarding battery life for a rechargeable battery is the number of times the battery is capable of being recharged. Batteries are not able to be operated through a charge and discharge cycle forever. The law of diminishing returns applies here.

Eventually, the output of the battery is reduced and the battery has to be replaced.

If a battery has a fairly low cost factor, this problem is not too important. When batteries become expensive, the charging cycle problem is apparent. The problem of cost as it relates to battery life has limited the use of lead-acid cells in the transportation field. Research is in process to overcome this problem. General Motors Corporation has announced a new development in this area. It is a cell constructed of a zinc-nickel oxide. This cell is able to store twice the energy as the conventional lead-acid wet cell. The cell also is able to be recharged more times than the standard cell. This reduces operational costs. It also goes a long way toward making the electric vehicle more feasible.

Thermal Cells

Heat, when applied under the correct set of conditions, can also produce electric energy. One device using the principle is the thermocouple. A thermocouple is illustrated in Figure 2-12. This device is

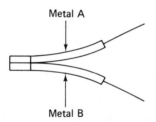

Metal A

Metal B

Figure 2-12. A thermocouple is constructed of two dissimilar pieces of metal. Their junction produces a voltage when it is heated.

made of two different kinds of metal strips joined at one end. The joined end is heated. When the metal is heated, a voltage develops between the opposite ends of the metal strips. This voltage has a very low value. Several sets of thermocouples may be joined in order to increase the developed voltage to a more useful level.

The thermocouple is often found as the core of a temperature-monitoring system. The industrial application is particularly good in devices such as industrial furnaces. The voltage produced by the thermocouple is directly proportional to the temperature at the junction of the two metal strips. Bimetal thermal cells are also used as sensors in heat-monitoring systems.

Photo and Solar Cells

Electrical energy may be produced from light sources. Certain chemical substances, such as sodium and potassium, will give off electrons when exposed to visible light rays. An application of this is shown in Figure 2-13. The device shown is called a *phototube*. It is a special

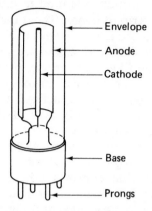

Figure 2-13. The photo tube produces an electron flow when light
strikes its cathode element.

kind of vacuum tube. The phototube has two elements. One is the cathode, which has a half-moon shape in this tube. The cathode is coated with a chemical that emits electrons when light rays shine on it. Immediately in front of the cathode is a rod-shaped element called the anode. Electrons emitted from the cathode strike the anode of the tube. When the tube is connected in a circuit such as the one shown in Figure 2-14, electrons will flow in the circuit. A sensing de-

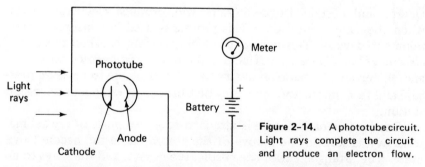

Figure 2-14. A phototube circuit.
Light rays complete the circuit
and produce an electron flow.

vice is connected into the circuit. When the electrons flow, the sensing device is activated. This could be used as an intrusion alarm. The alarm would sound when the electrons flowed in the circuit.

Another way to set up this circuit would be to have it always in an "on" condition. Then, if the light rays were to be interrupted, the sensing device would turn off. Turning off the sensing device would activate a second circuit, which would turn on an alarm system. In another application this circuit could be used as an industrial counter. Each time the light beam is interrupted a counter would trip. As items moved along a conveyor belt and past the light source, they would be counted.

Still another application of this principle is used in motion pictures. The film track is made up of a strip of bands. These bands relate to the frequency of the sound on the film. The bands are a variety of widths and shades of gray. These widths and shades are visual indications of the sounds recorded on the film. The varying intensity of a light beam as it tries to pass through the film and on to a phototube relate to the sound recording. This light beam is changed into a varying electric voltage by the action of the phototube. The varying voltage follows the sound waves originally recorded on the film. This voltage is amplified and changed back into sound waves by loudspeaker action in the motion picture projector amplifier.

The *photocell* is another device that converts light energy into electrical energy. Another name for this device is the *solar cell*. This cell is made of two layers of material. One layer is a light-sensitive material, such as silicon. The second layer is made of a different material. The two layers are arranged in such a way that light striking the silicon will force electrons onto the other metal. This process makes one plate positive and the other plate negative. The two plates, each with opposite electrical charges, form a cell.

Pressure Cells

Quartz and certain Rochelle salts will produce electrical energy when they are compressed. This process is used in consumer entertainment devices. This voltage-producing system is called the *piezoelectric effect*. These devices include the phonograph, color television, and citizen's band radio. Industry uses the principle to measure pressure and to maintain the frequency of a broadcast radio and television station.

The phonograph needle is attached to a thin piece of crystal material. Variations in the groove of the record cause the needle to vibrate. The arrangement of the needle, the quartz, and the wires to an external circuit is shown in Figure 2-15. The vibration of the needle

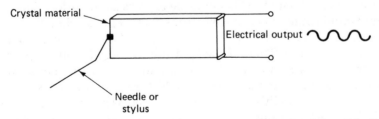

Figure 2-15. A crystal used as a phonograph pickup. Movement of the needle distorts the crystal and produces a voltage at the output connections.

makes the quartz piece change its shape. This produces a voltage across the wires to the external circuit. The voltage corresponds to the information on the record. The electrical variations produced are connected to an amplifier and its loudspeaker. The loudspeaker converts the electrical waves into sound waves that can be heard by the human ear.

A variation of this principle is used in color television and all types of radio transmitting equipment. If a voltage is connected to a piece of quartz crystal material, it will vibrate. The vibrations occur at a very high rate. This produces a voltage that varies at a constant rate. This voltage is used to establish specific operating frequencies for radio and television devices. The frequencies produced by the voltage are constant. The frequency (or rate of vibration) is determined by the size and shape of the crystal material as well as the amount of voltage used to get it to vibrate.

CELLS IN COMBINATION

Each electrical energy source described in this chapter produces a limited amount of electrical energy. In many cases these amounts are so low as to make the cell not practical. Individual cells may be wired together to have a larger amount of energy available.

There are two basic ways of wiring cells together. One way is illustrated in Figure 2-16. This is called a series connection. When cells

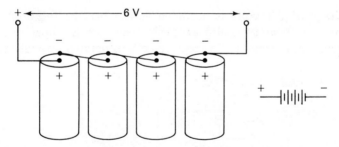

Figure 2-16. Individual cells may be connected in series as shown. Voltage produced by each cell adds to provide a battery voltage.

are wired in series, each individual-cell voltage is added in order to determine the total voltage at the terminals of the package. In the case of a dry or wet cell, the total package is called a *battery*. The battery has two terminals. These are usually the only connection from the battery to the work load. Each cell is connected inside the battery to produce the terminal voltage. Automobile batteries contain six cells. Each cell produces 2.1 volts. Adding the voltage from each cell gives a terminal voltage of 12.6 volts.

The current-producing capability of the series-wired battery is limited to the current produced by a single cell. This is due to the way in which the cells are wired and to the way in which the current flows through the battery. Current flow must travel through each cell in turn in order to get from one terminal in the battery to the other terminal. The maximum current flow is limited to the capability of the smallest cell in the chain. If six 2.1-volt cells, each with a capacity of 1.5 amperes, were wired in series, the terminal voltage would be 12.6 volts. The current capacity of the battery would be limited to 1.5 amperes in this circuit. If one cell in this battery were to fail, the whole battery would fail.

A second way to wire cells is in parallel. This is shown in Figure 2-17. Here each cell is producing its terminal voltage of 2.1 volts. The

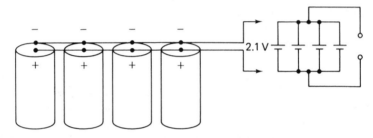

Figure 2-17.   Individual cells may be connected in parallel as shown.   In this configuration the current capability is increased to the sum of each cell's current rating.

cell voltage of 2.1 volts is available at the battery terminals. Each cell's current capacity is also available. The current capacity adds in this circuit arrangement. Connecting cells in parallel provides the ter-

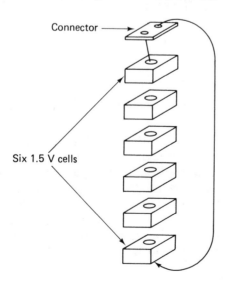

Connector

Six 1.5 V cells

Figure 2-18.   The 9-volt transistor radio battery is constructed of six 1.5-volt cells that are series connected.

minal voltage of one cell, but a total of each cell's current capability. If each cell had a capability of producing 1.5 amperes of current and six cells were wired in parallel, then the battery would produce 2.1 volts at a maximum current of 9 amperes.

Batteries used in the radios and calculators use these same types of circuits. The common 9-volt battery uses six 1.5-volt cells wired in a series configuration. The six cells' current is limited to the maximum current capability of each single cell. If one cell has a reduced voltage or current capability, the performance of the battery is also limited. The construction of this battery is shown in Figure 2-18.

ELECTROMAGNETIC PRODUCTION

One of the most common ways of producing large amounts of electric power is by means of the electromechanical generator. This topic is very broad and complex. It will, therefore, be presented in full detail in Chapter 9.

SUMMARY

Electric power is produced in several different ways. To understand how power is produced and used, a person first must know the technical terms and phrases commonly used. Terms and phrases presented in this chapter include voltage, EMF, current, direct current, alternating current, resistance and power. Each term is defined as follows:

Voltage is a difference in electrical pressure between two points in any circuit. Voltage is the electrical force that makes electrons move in a circuit.

EMF means electromotive force. This is another term for electrical pressure. EMF is also measured in volts.

Current is the movement of electrons in a circuit. Electron movement is measured in amperes. Electron current flow is from negative to positive in a circuit.

Direct current, or dc, is the flow of electrons in a circuit in one direction at all times.

Alternating current, or ac, is a flow that periodically reverses itself in the circuit. The rate of one complete alternation is called the frequency of the current.

Resistance is the measured opposition to the electron flow in a circuit. The ohm is the unit used to measure resistance.

Power is the time rate of doing work in a circuit. Electric power is measured in units of the watt.

Material related to scientific notation is also included in this chapter. Prefixes used by technical personnel to replace large values include kilo- and mega. Prefixes in common usage to replace small values are milli-, micro-, and pico-. Kilo- replaces 1,000; mega- replaces 1,000,000. Milli- is 1/1,000, micro- is 1/1,000,000, and pico is used to replace 1/1,000,000,000,000. This saves using a lot of zeros and also tends to minimize mistakes when reading a quantity. These terms may be used to describe the quantities of any voltage, resistance, current, or power value.

Electrical energy is produced in several ways. These include chemical cells, thermal cells, photo cells, and pressure cells. Chemical cells are either wet or dry. The wet cell uses a liquid electrolyte. The cell produces about 2.1 volts. It is rechargeable. Dry cells may be made of combinations of several materials. These include lead and zinc, manganese dioxide and zinc, mercuric oxide and mercury, and nickel and cadmium. Most of these cells produce about 1.5 volts. All except the nickel-cadmium cell are not normally rechargeable.

Thermal cells are made from two dissimilar metals. Two strips of these metals are connected at one end. This end is heated. A voltage is produced at the "cold" end of the strips.

Photo and solar cells use light sources to produce electrical energy. Light striking certain chemical surfaces will cause a flow of electrons. The electron flow produces a voltage between two plates in the cell. This voltage is used to power an external circuit and to perform work.

Quartz and some other crystal materials are capable of producing a voltage when pressure is applied to the crystal material. This principle is used in phonograph pickups and microphones. If a voltage is applied to a piece of quartz, it will vibrate at a specific frequency. This principle is used in radio and television sets in order to permit operation on one frequency.

Cells may be connected in series or parallel. This combination of cells is called a battery. A series connection will produce a voltage that is the sum of the voltages of each individual cell. The current is limited to the current of one cell in this configuration. Cells wired in parallel will produce currents that are the sum of each cell's current capability. Voltages in a parallel-wired connection are limited to the voltage of one cell.

QUESTIONS

2-1. The fluid power system requires

    a. pump, lines, and storage

    b. lines, pump, and source

    c. pump, lines, and load

    d. lines, load, and storage

2-2. Electrical voltage is

   a. electron flow

   b. electromotive force

   c. opposition to electron flow

   d. none of the above

2-3. Electromotive force is measured in units of the

   a. ampere

   b. ohm

   c. coulomb

   d. volt

2-4. In electrical circuits the work unit is called the

   a. opposition

   b. resistance

   c. force

   d. load

2-5. Chemical cells are classified as

   a. wet cells

   b. dry cells

   c. both of the above

   d. none of the above

2-6. Dry cells

   a. are rechargeable.

   b. are nonrechargeable.

   c. Charging depends upon the chemical makeup of the cell.

   d. Charging is independent of the chemical makeup of the cell.

2-7. Cells use ____ action in order to produce energy.

   a. chemical

   b. thermal

   c. light

   d. pressure

   e. all of the above

2-8. Six 1.5 volt, 2-ampere cells series wired as a battery will produce

   a. 1.5 volts at 2 amperes

   b. 3.0 volts at 9 amperes

   c. 4.5 volts at 6 amperes

   d. 9.0 volts at 2 amperes

2-9. Six 1.5-volt, 2-ampere rated cells wired in parallel will produce

    a. 1.5 volts at 2 amperes

    b. 1.5 volts at 12 amperes

    c. 9 volts at 6 amperes

    d. 4.5 volts at 2 amperes

2-10. Cells may be wired in

    a. series

    b. parallel

    c. both of the above

    d. neither of the above

2-11. Match the terms on the left to the proper description given on the right.

1. ampere	a. cycles per second
2. volt	b. pressure
3. ohm	c. 1,000
4. watt	d. electron flow
5. hertz	e. 1/1,000,000
6. load	f. 1,000,000
7. source	g. work
8. kilo-	h. opposition
9. mega-	i. energy
10. milli-	j. 1/1,000
11. pico-	k. power

3

Resistance and Resistors

Three basic factors affect the flow of electrons in any electrical circuit. These are capacitance, inductance, and resistance. The counter EMF produced by inductive action tends to control the electron flow in any circuit.

Capacitance is a force that will work to control the voltage levels in an electrical circuit. The concept of capacitance and its relation to electron flow is covered in Chapter 11.

The third factor affecting circuit current is resistance. This is probably the easiest of the three factors to understand. This is because resistance in a circuit is a fixed factor. It is not related to time or to the amplitude of other circuit values. There are quite a few factors involved in the description of any resistor. These relate to size, shape, and power handling, as well as to the materials from which they are manufactured. Each factor is discussed in the following paragraphs.

RESISTANCE

Electrical resistance refers to the opposition of the flow of electrons in an electrical circuit. All materials known to science have some amount of electrical resistance. The ability to conduct electrical current depends upon the quantity of free electrons in any substance. The free electrons are based on the atomic structure of the material. Those materials having many free electrons are considered to have a low resistance. Electron flow through these substances is relatively high. These materials include copper, gold, silver, and aluminum. Keep in mind that an electromotive force (EMF) is required to move electrons in an electrical circuit. Electrons do not have direction or any quantitative values without an applied EMF. Resistance refers to the ease with which electrons may move under circuit conditions.

Materials that have few free electrons in their atomic structure are called nonconductors. Sometimes they are called *insulators*. Materials in this group include glass, rubber, most plastics, and mica. There are exceptions to every scientific rule. This also holds true for conductors and nonconductors. Scientists have produced glass sheets that will conduct electricity. These, however, are not typical and should be categorized as unusual. This book will present the normal situation rather than the exceptions.

A third group of materials are neither conductors nor nonconductors. This group falls between the two extremes. They are called semiconductors. They allow electron flow under specific conditions. Semiconductor materials and their application are commonly seen by electrical technicians in the form of diode rectifiers, transistors, and integrated circuits. Semiconductor theory and application is discussed in Chapter 16.

Units of Resistance

The basic use of resistance is the ohm. It is represented by the Greek letter Ω (omega). The ohm is defined as the amount of resistance that permits 1 ampere of electrical current to flow in a circuit when 1 volt of EMF is applied to the circuit. These conditions will produce 1 ohm of resistance. The term ohm when used to describe resistance refers to the amount of resistance in an electrical circuit. Some numerical value appears with the ohm. The number refers to the specific quantity of resistance. The phrase, "there are 200 ohms of resistance in this circuit," is typical of those used to describe resistance.

Large amounts of resistance use prefixes to describe the ohmic value. Quantities in the range between 1,000 and 999,000 use the prefix *kilo-* to represent the last three zeros. Accepted usage is 24 kilohm, or 24 kΩ, instead of twenty-four thousand ohms. A second prefix is used to represent quantities of 1 million or more. This term is *mega*, or m. High-value resistors are described as 10 megohms or 10 mΩ, instead of ten million ohms. Very small quantities of resistance, such as those under 1 ohm, are not often used in electrical circuits. When they are used, they are described in decimal terms. A low resistance value of 0.01 ohm is the acceptable manner for presenting these values.

The letter R is used to represent a specific resistance. Often there will be a numerical subscript with the letter. For example, R_{21} will be used to identify a specific resistor on an electrical schematic diagram or blueprint. Elsewhere on the drawing, or on a separate list of parts, this resistor will be described in detail. The reason for this is to help keep the schematic diagram readable. In many cases, if all the relevant information about every part on the schematic diagram was printed on the diagram, there would be little or no room left to show the wiring. Use of letters and subscripts makes for a cleaner diagram. These are easier to read and may also be smaller in size than those containing every possible bit of information.

CONDUCTANCE

In some cases it may be best to describe the ease in which electron flow occurs in a circuit. When this is required, the electron flow is described in units of conductance rather than resistance. Conductance is the reciprical of resistance. The letter G is used to represent conductance. The unit of conductance is the siemens. The letter S represents the term siemens. The formula for conductance is:

$$G = \frac{1}{R}$$

where G is in siemens and R is in ohms.

TYPES OF RESISTORS

Resistors are categorized in several different ways. Each adds a certain dimension to the total description. When each category is described, then, and only then, is the description complete. Material in this section describes the several categories in which resistors fall.

Ohmic Value

Ohmic value relates to the amount of resistance in a specific resistor. Each resistor produced has a design value of resistance. Very often, resistors are manufactured that vary from the design value. Each resistor type has a tolerance rating. This may vary anywhere from 0.5% to 20%. This means that the exact value of resistance of the specific resistor may not be as marked on the resistor. A 100-ohm resistor with a ± 10% tolerance may range between 90 and 110 ohms in value and still be within the tolerance range. This resistor is acceptable for use in a circuit where the tolerance is not critical. Tolerance does not mean that the resistor is off from its design value by 10% or so. All it means is that it *could* be off and still be acceptable. The only way to be certain is to measure the ohmic value with a resistance meter. Conversely, a resistor having a ± 20% tolerance whose ohmic value was exactly 100 ohms could be used to replace a 100-ohm ± 1% resistor successfully. Generally, the closer tolerance resistors are more expensive than those with high tolerances.

Power Rating

Resistors are also described by their ability to handle electrical power. This rating is in watts. It gives the maximum power rating for the resistor. Additional amounts of power tend to destroy the resistor. The lowest standard rating is $\frac{1}{8}$ watt. The physical size of the resistor container is related to its power-handling rating. As the power rating increases, so does the physical dimension of the package.

A group of low-power resistors is shown in Figure 3-1. The ratings

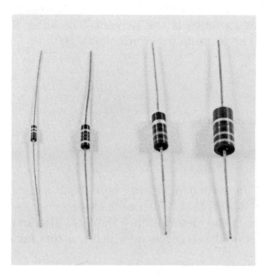

Figure 3-1. Low-power resistors range from $\frac{1}{8}$ to 2 watts in size. Each has its specific physical characteristics, with the $\frac{1}{8}$ -watt resistor being the smallest in size.

go from a low $\frac{1}{8}$ watt to a high of 2 watts. The physical size of each wattage rating has become fairly standard in the industry. Identification of wattage ratings by physical size is commonplace. There is a break between low- and high-power resistors at the 2-watt point. Resistors rated at 3 watts or more are considered to be power resistors. Their construction is different than their low-power counterparts. Some of these are shown in Figure 3-2. Here, too, size is a good indicator for the power-handling capability of the resistor.

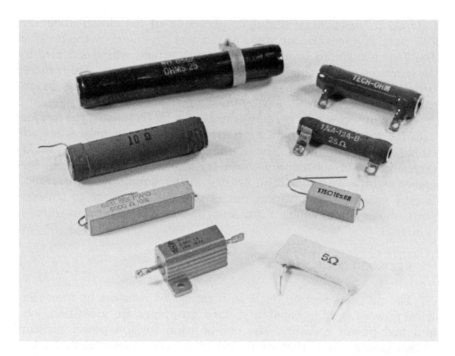

Figure 3-2. Power resistors are larger is physical size. Here, too, the power rating determines the specific size.

There is a middle-ground group of low-power resistors that are capable of handling 3 to 4 watts of power. These are shown in Figure 3-3. These resistors have a metal housing. The housing is securely

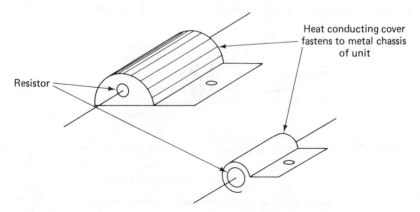

Resistor

Heat conducting cover fastens to metal chassis of unit

Figure 3-3. Low-power resistors may have a heat-dissipating device in order to handle higher power requirements.

fastened to sheet metal during construction of the electrical circuit.
The metal surface takes on some of the heat produced by the resistor.
The heat is diffused over the sheet metal surface, which is cooled by
the surrounding air. This is called *heat sinking* the device. This process
permits higher wattage use than the basic rating of the device. In any
case, too much heat will destroy the device.

In many cases a larger power resistor may be substituted in a circuit for a lower power rating resistor. This is true only when the
circuit design permits it. The ohmic value of both resistors must be
the same. Never substitute a lower wattage rated resistor for one of
higher ratings. The substitute resistor in this situation will overheat.
It may catch on fire, causing circuit damage or damage to adjacent
parts, the device in which it is placed, or even the building in which it
is housed.

RESISTOR COMPOSITION

The previous section deals with ratings of resistors by their elec-
trical-power-handling capability. Two groupings are identified, low-
and high-power resistors. The composition for low-power resistors
differs greatly from that of their high-power counterparts. Each is
discussed in this section.

Low-Power Resistors

Low-power resistors are usually called *composition* resistors because
the standard material for low-wattage resistors is a carbon composite
material. A cutaway drawing for a composition resistor is shown in
Figure 3-4. Note that the resistor element is located in the center of

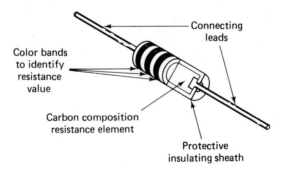

Figure 3-4.   Construction of a carbon composition resistor.

the housing. Wire leads are attached at each end of the resistor element. These leads connect the resistor element to the circuit in
which the resistor is used.

For many years the resistor element was made from a combination of materials. The basic resistive material used in this combination is carbon; hence the name *carbon composition* resistor. This name is often shortened to *carbon* resistor. The ratio of carbon to the other materials determines the specific ohmic value of any carbon resistor. Ohmic values for carbon resistors will range from 1 ohm up to 20 or more megohms. The physical size of carbon resistors has nothing to do with the ohmic value of the resistor. Physical size of any resistor basically relates to its power-handling capability.

Another type of resistor finding increased application in industry is the *cermet* resistor. This resistor uses a metallized ceramic material instead of carbon composition for its resistive element. The reason for the use of this resistor is the higher reliability of the wire lead connection to the resistance element. Cermet resistors also have closer tolerance to the ohmic value than do carbon composition resistors. A drawing of a typical cermet resistor is shown in Figure 3-5.

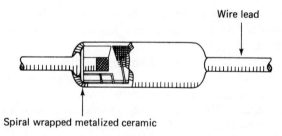

Figure 3-5. Construction of a metallized ceramic resistor.

The cermet resistance element is spirally wound around an aluminum core material. The wire leads are connected to the cermet element. The ohmic value of this type of resistor depends wholly on the size of the cermet element.

Metallized film materials, similar to the cermet resistor, are also used to make resistors. The advantage of this type of resistor is its physical reliability. It, too, is produced with a deviation from marked value of less than 1%. Both the film resistors and cermet resistors are more expensive than comparable carbon composition resistors.

High-Power Resistors

Carbon resistors have one drawback. This is the ability of the resistor to handle large amounts of electrical power. Most carbon- and film-type resistors are limited to low-power applications. When large amounts of electrical power are handled, the resistor element usually is made with a resistance wire. This leads to the classification of *wire wound* for high-power resistors. An illustration of this type of resistor is shown in Figure 3-6. The wire most often used for this type of resistor is a nickel-chrome alloy called nichrome wire. The amount of

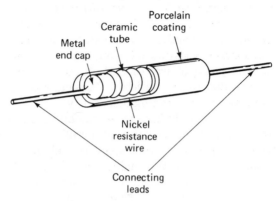

Figure 3-6.   Construction of a wire-wound resistor.

resistance in any given piece of wire is determined by its diameter, length, and resistivity. Nichrome wire has a fairly high amount of resistance per foot. Therefore, it is easily used when making wire-wound resistors. The amount of power handling is directly proportional to the physical size of the resistor. As power-handling requirements increase, the physical size of the resistor also increases. Resistors with high power ratings in the order of 200 watts may have a diameter of $1\frac{1}{2}$ to 2 inches and a length of 6 or 8 inches. Most of these power resistors are wound on a ceramic core form. The core is hollow. This allows more air circulation and thus, more cooling, as the power resistor dissipates its heat.

RESISTOR MARKINGS

The point was made that all resistors of a specific power rating have approximately the same size. The logical question now is, "How does one tell the ohmic value of a resistor?" The answer is that all resistors have markings on them to indicate the ohmic value. In some special cases, resistors are marked with the manufacturer's part number rather than the ohmic value. This most often happens with power resistors rather than the carbon composition types.

Power resistor values are easy to read. The manufacturer simply prints the ohmic value on the side of the resistor. Normally, a high-contrast ink color is used. An example is a white or silver ink on a brown resistor body. The print size is fairly large on these resistor type, making them easy to read.

Low-power resistors use another type of marking system to identify their ohmic value. This system uses several bands of colors. The colors are related to the Electronic Industries Association color and number system. This system, which relates specific colors to numbers, is shown in Figure 3-7. The colors run from black to white, using the

Black	0	Blue	6
Brown	1	Violet	7
Red	2	Gray	8
Orange	3	White	9
Yellow	4		
Green	5		

Figure 3-7. The Electronic Industries Association standard color code.

colors of a rainbow as a reference. This type of color-numbering system is also used in multiwire cables to identify specific wires. It should be learned by all who are working in the field of electricity and electronics.

Low-power composition-type resistors use the color code system, also. Each resistor has several bands of color on it. The number of color bands will vary between three and five. Each band represents a specific part of the marking system. This system is illustrated in Figure 3-8. It is the industry-standard color-code system for low-power resistors.

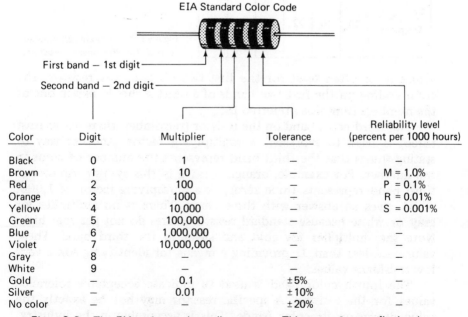

EIA Standard Color Code

First band — 1st digit

Second band — 2nd digit

Color	Digit	Multiplier	Tolerance	Reliability level (percent per 1000 hours)
Black	0	1	—	—
Brown	1	10	—	M = 1.0%
Red	2	100	—	P = 0.1%
Orange	3	1000	—	R = 0.01%
Yellow	4	10,000	—	S = 0.001%
Green	5	100,000	—	—
Blue	6	1,000,000	—	—
Violet	7	10,000,000	—	—
Gray	8	—	—	—
White	9	—	—	—
Gold	—	0.1	±5%	—
Silver	—	0.01	±10%	—
No color	—	—	±20%	—

Figure 3-8. The EIA resistor color-coding system. This uses a four- or five-band color-coding system; the colors represent numbers, multipliers, or tolerances.

Color bands always start at one end of a resistor. If the first band is not exactly at the resistor's end, it is closer to one end than to the other end of the resistor. The first band always represents a whole number. It also represents the first significant digit of the numerical value of the resistor. The first color band will never start with zero in this system. If you are reading a resistor color code and see a black band as the first color band, then reverse the resistor or start reading the colors from the opposite end. The second color band also represents a whole number, or the second significant digit. This, also, will always be a number. The electronic industry has a preferred number system for resistor values. This system, shown in Figure 3-9, gives the

EIA Preferred Number Series

Preferred number series						
±5% Tolerance	10 11 12 13	15 16 18 20	22 24 27 30	33 36 39 43	47 51 56 62	68 75 82 91
±10% Tolerance	10 12	15 18	22 27	33 39	47 56	68 82
±20% Tolerance	10	15	22	33	47	68

Figure 3-9. A preferred number system for low-power resistors.

values most often used for the first two digits of any resistor. The color coding on the first two bands of a resistor should match one of the numbers from this preferred list.

The third color band on the resistor (remember, there are at least three) is used to represent a multiplying factor. Another way of stating this is that the third band represents the number of zeros in the number. For example, orange is three in this system. An orange third band represents three zeros, or a multiplying factor of 1,000, which gives an answer with three zeros. There is no multiplier for gray or white because standard resistor values do not get that high. Note the multiplier for gold and silver in the third band. These values are less than 1, providing a means for identifying some very low resistance values.

The fourth color band is used to indicate acceptable tolerance values for the resistor. A specific resistor may not be exactly the value shown on its color bands. This is acceptable in the industry.

The tolerance band provides information that gives some leeway for these values. A 10% tolerance value indicates that the resistor's exact value may be as low as a point that is 10% lower than its marked value. Or it may be as high as 10% above the marked value. It may fall anywhere between these two extremes and still be acceptable for use in a circuit. If the fourth color band is not present, the resistor tolerance is ±20%. In any case, a 10% or 20% tolerance resistor may be used in place of a 5% tolerance resistor if its measured value is in the 5% range. Note the three groups of values in the preferred number series. This shows which resistor values are available in each tolerance range. This information should be helpful when identifying or ordering low-power resistors.

The final color band on a resistor is a recent addition to the series. This band represents a level of reliability. It specifies how many resistors from a specific production run will fail after 1,000 hours in a test circuit. This information is not found on all resistors. It is a result of high reliability demands due to space, military, and heavy industrial use.

Some examples of how the resistor color code system works follow:

Read the color code values for each resistor:

red,   red,   yellow,   silver
(2   2   0,000   ±10%)

brown,   black,   red,   gold,   red
(1   0   00   ±5%   0.01% reliability)

green,   blue,   orange
(5   6   000   ±20%)

Convert these values into a color code:

10,000 ± 10%   0.001 reliability
(brown, black, orange, silver, orange)

2,700 ± 5%
(red, violet, red, gold)

VARIABLE RESISTORS

The types of resistors discussed to this point are called *fixed* resistors. Their ohmic value is set during the manufacturing process. It cannot be changed under normal operating conditions. There are many requirements for resistors with a variable resistance factor.

These resistors are manufactured in a variety of shapes. There are, however, some groupings for these resistors. Keep in mind when reading this section that the illustrations show only some of the shapes. There are many others, too numerous to show, that are also in use. As you work with these devices, you will become familiar with the shapes used in the equipment you use.

A drawing of the parts of a variable resistor is shown in Figure 3-10. This unit has a wire-wound resistance element. The wire is

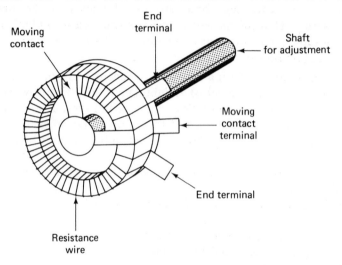

Figure 3-10.   A variable resistor has a slider that permits adjustment over a fixed range of resistance.

wound around a ceramic form. It is connected to the two end terminals of the device. In some instances, a carbon element is substituted for the wire-wound element. The classification of either wire-wound or composition is made from this construction. The carbon element is flat and U shaped. It fits in the space occupied by the wire-wound element. A third connection is made to the variable resistor. This connection is to a moving contact, which slides on the resistor element. The element is also attached to a shaft. The shaft rotates about 300 degrees. This produces a variable resistor.

Figure 3-11 shows the electronic symbol for both a fixed resistor

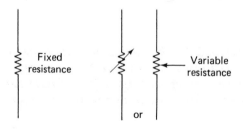

Figure 3-11.   Electronic symbols used to represent variable resistors and fixed resistors.

and a variable resistor. The symbol for a resistor is standard. It is shown on the left side of the drawing. The variable resistor has a third connection, which is the part that is variable. The arrow indicates that the device is variable by means of a moving shaft connection. There are two symbols given for a variable resistor. The one on the left indicates a type of variable resistor called a potentiometer, or just *pot*. An explanation of how this works is shown in Figure 3-12. The numbers 1, 2, and 3 represent the terminals of the device. The total resistance is always found between the ends of the resistance element. A certain amount of resistance is present between one end terminal 1 and the variable terminal 2. This is less than the total resistance of the device. Identify this as resistance A. A certain

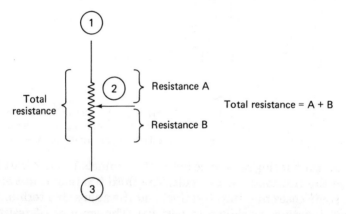

Figure 3-12. The variable resistor has a fixed total value of resistance between its outer terminals. The arm is used to vary the resistance between it and one of the outer terminals.

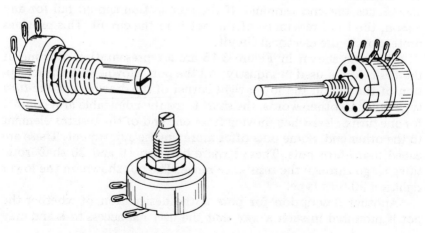

Figure 3-13. Shapes of typical potentiometers.

amount of resistance is also present between the adjustable terminal 2 and the other end terminal 3. Identify this as resistance *B*. The exact amount of resistance between terminal 2 and the ends will vary depending on the position of the adjustable arm of the pot. In any case, resistance *A* and resistance *B* will add to equal the total resistance between the adjustable arm and the ends. Drawings of typical pots are shown in Figure 3-13.

A second way of using the same pot is to only utilize two of the three terminals. One end terminal and the adjustable terminal are used. This is illustrated schematically in Figure 3-14. In this type of

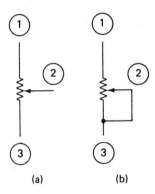

(a)   (b)

Figure 3-14. Use of two of the three terminals is possible. This is called a rheostat connection.

connection the varying resistance between terminals 1 and 2 is utilized to change the resistance in a circuit. The third terminal is not always used. In some cases the third terminal and the adjustable terminal are connected together, as shown in part (b). The amount of resistance between terminals 1 and 2 still varies as the control shaft is turned. The resistance between these terminals decreases as the movable arm approaches the end terminal. If the arm section should fail for any reason, the total resistance of the pot is in the circuit. This provides protection to the electrical circuit.

The pots shown in Figure 3-15 are a representation of some of the basic forms used in industry. All the pots shown, with the exception of the one in the lower right corner of the drawing are *one-turn* type pots. In other words, the shaft turns the adjustable arm contact for one turn or less when moving from one end of the resistor element to the other end. Some pots offer more precise adjustment. These are called *multi-turn* pots. These require between 10 and 20 shaft rotations to go through the resistance range. The pot shown on the lower right is a 20-turn type.

Another description for pots is the designation of whether the pot is mounted in such a way that the user has access to it and may adjust it. Many pots are mounted inside control panels or on printed circuit boards. These are used to make minor adjustments in circuit values. In a sense they "trim up" the circuit. Because of this they are called *trim pots*.

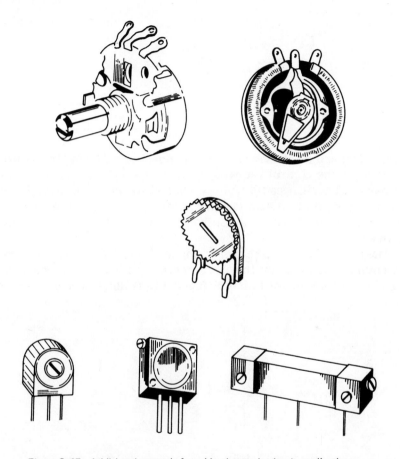

Figure 3-15. Additional controls found in electronic circuit applications.

POTENTIOMETERS AND RHEOSTATS

The variable resistor may be wired into a circuit in one of two ways. One method varies the voltage across two of the terminals of the device. When a variable resistor is used in this way, it is called a potentiometer. It meters, or controls, the potential in the circuit.

This is shown in Figure 3-16. The applied voltage is connected between terminals 1 and 3. This is called the circuit input voltage.

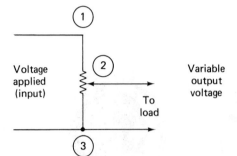

Voltage applied (input)

To load

Variable output voltage

Figure 3-16. The variable resistor connected as a potentiometer, or potential adjuster.

Terminals 2 and 3 are used as output connections. Moving the adjustable arm of the control varies, or meters, the output voltage, hence the name. This changes the voltage applied to the work load. These circuits are used to control electronic signals. We know them best as volume controls, brightness controls, or contrast controls in television sets.

The second circuit is shown in Figure 3-17. This is called a rheostat circuit. In this circuit the electron current is controlled when the variable resistor arm position is changed. Controlling electron current

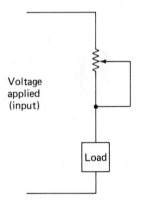

Voltage applied (input)

Load

Figure 3-17. An application of a variable resistor connected as a rheostat.

makes motors change speed, lamps dim, and provides information as to the fuel level of an automobile gas tank.

ADJUSTABLE RESISTORS

Some power resistors are adjustable. Most of these have an adjustment arrangement that is very different than their variable counterparts. One such type is shown in Figure 3-18. Part (a) shows a typical power resistor. The wire-wound element is covered with a ceramic

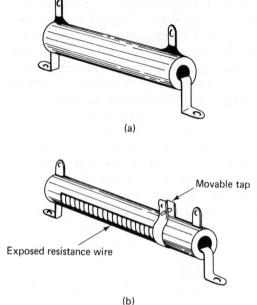

(a)

Movable tap

Exposed resistance wire

(b)

Figure 3-18. Adjustable resistors are wire-wound devices. They have a means of adjusting the overall resistance value.

coating. None of the wires is exposed. Part (b) shows how this type of resistor is made into an adjustable type. Part of the wire-wound element is left exposed during the manufacturing process. A metal ring circles the resistor body. Part of the ring is indented so that it will ride against the resistor element. The ring is held in place by a nut and bolt that clamp the ends of the ring together. The ring is adjustable, but once in place it is usually left in that position. This type of resistor is called an adjustable power resistor. Its symbol differs from the variable type by the use of a dot instead of an arrowhead to show the connection point.

TESTING RESISTORS

Resistors do fail. In many cases the failure is caused by some other component in the circuit. The breakdown of the resistor is the result rather than the cause for failure. There are really only two ways in which a resistor may fail. One way is the failure of the wire connection between the resistor element and the electrical circuit to which it is connected. The second way is usually caused by excessive heat. The heat is produced when an excessive amount of current flows through the resistor. This produces heat in the resistive element. The events that follow the buildup of heat are heating, charring, and cracking of the resistor housing. This is followed by an increase in the amount of resistance. The increase may be a small amount, only sufficient to make the resistor go out of tolerance. In extreme cases the heat literally "opens" the resistance element. When this occurs, the element has an infinite amount of resistance. This resistance is too high to measure. A very high resistance such as this will stop circuit current flow, which stops the circuit from working properly.

Resistors are tested with a device called an ohmmeter, a special type of tester that is designed to measure resistance. To make an accurate resistance measurement, at least one end of the resistor must be disconnected from the circuit in which it is wired. The leads from the ohmmeter are connected one to each side of the resistor. The resistor value is then read from the meter dial.

Carbon resistors also have a voltage tolerance based on the wattage rating of the resistor. This range is shown in Figure 3-19. There are

Power rating (watts)	0.125	0.25	0.5	1.0	2.0
Working voltage rating (maximum value)	150	250	350	500	750

Figure 3-19.   Voltage ratings of low-power resistors.

two ways of determining this rating. One is to take the square root of the product of the wattage rating and the ohmic value of the resistor. This will give a specific voltage value. The second way is to use general values of voltage for each wattage rating. Both are given in this illustration. Exceeding the voltage rating may cause an arcing condition inside the resistor body. This will produce a change in the value of resistance of the resistor.

ODDS AND ENDS

Some variable resistors are *gang* mounted. This is accomplished in one of two ways. In the first way the unit has one shaft. All the rotating slider arms are connected to this shaft. All rotate at the same time. This is used in devices where two similar adjustments are neces-

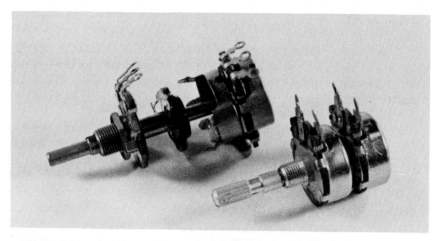

Figure 3-20. Variable resistors may be "gang mounted" in one of two ways. These are either with one common shaft or use of two concentric shafts.

sary, such as the tone control on a stereo unit. These are shown in Figure 3-20.

The second type of ganged control has two shafts. One shaft is hollow. It connects to the front control. The second shaft has a narrower diameter than the first. It fits inside the hollow shaft and connects to the back control. Each shaft may be rotated independently of the other. This type of control is called a *concentric* control. It is used when two different settings are required. A control of this type saves panel space. One common use is the contrast and brightness control on certain television sets.

Some resistors are designed to react to changes in temperature. This effect is used successfully in the *thermistor*. The thermistor's ohmic value changes when an electron current heats it. The thermistor may have either a positive or a negative temperature reaction. It may increase in resistance in some applications. In other applications the thermistor decreases in resistance. Selection of the proper type of thermistor is a design factor.

Studies of failures and fires in some television sets have led to the development of a flameproof type of resistor. Prior to the use of this type of resistor there were cases where resistors actually started to burn as they failed. Always replace a flameproof resistor that has failed with another flameproof resistor. Failure to do so could lead to a fire in the set.

SUMMARY

A significant factor in any electrical circuit is resistance. Resistance is the opposition to the movement of electrons in the circuit. There are three groups of materials in electrical work. These are classified as conductors, semiconductors, and insulators. Each has its place in the study of electricity.

The unit of measure for resistance is the ohm. One ohm of resistance is encountered when 1 volt creates 1 ampere of current flow in a circuit. The prefix kilo- is used to describe thousands of ohms. The prefix mega- is used to describe millions of ohms of resistance. The letter R indicates a resistance. Subscripts identify specific resistors in a system. For example, R_{41} identifies a specific resistor. A parts list is required to find the specific values of this resistor.

Conductance is the reciprical of resistance. It is measured in units of the siemens. The letter G is used to represent conductance values.

Resistors are categorized by their ohmic value and their power-handling ratings. Common power ratings range from $\frac{1}{8}$ watt to over 200 watts. The power rating is used to tell how much electrical power may safely be converted into heat without damage to the resistor. Resistors rated at less than 3 watts are considered to be low-power types. Any resistor with a power rating of 3 watts or more is called a power resistor.

Low-power resistors are made of a carbon composition material. They are often called *carbon* or *composition* resistors. Some of these types are made of metallized ceramic material. They are called *cermet* resistors. Another group is made by depositing a metallized film on a ceramic form. These are called *film* resistors. Power resistors are usually made by winding a resistance wire on to a ceramic form. They are often called *wire-wound* resistors.

Low-power resistors use a color-band coding system to identify their ohmic value. Power resistors have the actual numeric ohmic value printed on them. If the ohmic value cannot be determined with either of these methods, a resistance-measuring device called an *ohmmeter* is used.

Resistors are made as either fixed-value or variable-value devices. Variable resistors have a means of adjusting their ohmic value. One type is a three-terminal device. Two terminals are connected to the total resistance value. The third terminal is a moving contact. It is adjusted by rotating a shaft that is connected to the contact arm. These types of variable resistors are called *potentiometers* or *rheostats*.

Power resistors are often produced with part of the wire winding exposed. A slider and clamp are placed around the body of the resistor. These devices are called adjustable power resistors. They function in a manner that is similar to the variable resistor, except that the contact on the slider is clamped at that point once it is established. It is not as convenient to change its position as with the variable resistor.

Special resistors are also found in electrical circuits. These include temperature- and voltage-sensitive resistors. Also included in some consumer products is a group of flameproof resistors.

PROBLEMS

Determine the color codes for the following resistor values.

3-1.   1 kΩ ± 10%

3-2.   4.7 kΩ ± 5%

3-3.   33 kΩ ± 5%

3-4.   68 kΩ ± 10%

3-5.   100 kΩ ± 20%

3-6.   1.2 MΩ ± 10%

3-7.   39 MΩ ± 20%

Identify the following resistor color coding.

3-8.   Red, red, yellow

3-9.   Brown, black, brown, gold, red

3-10. Violet, green, silver, gold

3-11. Orange, white, black, gold

3-12. Gray, red, red, gold

3-13. Orange, orange, orange, silver

3-14. Red, violet, yellow

QUESTIONS

3-1.   Opposition to the movement of electrons in a circuit is called

   a. conductance

   b. electrical opposition

   c. current retardation

   d. resistance

3-2.   The ease with which electrons flow in a circuit is called

   a. conductance

    b. electrical flow

    c. current movement

    d. resistance

3-3. Electrical resistors are rated in

    a. power-handling capacity

    b. ohmic values

    c. conductance values

    d. a and b above

    e. none of the above

3-4. Composition resistors are used in

    a. low-power circuits

    b. high-power circuits

    c. mixed power circuits

    d. all of the above

3-5. Wire-wound resistors are used in

    a. low-power circuits

    b. high-power circuits

    c. mixed power circuits

    d. all of the above

3-6. There are at least ____ color bands on low-power resistors.

    a. 2

    b. 3

    c. 4

    d. 5

3-7. On any resistor, the third color band designates

    a. the first significant number

    b. the second significant number

    c. the tolerance rating

    d. the multiplier

3.8. Resistors are available as

    a. fixed units

    b. adjustable units

    c. variable units

    d. all of the above

3-9. A thermistor is a

    a. voltage-sensitive resistor

    b. environmental resistor

    c. heat-sensitive resistor

    d. cold-sensitive resistor

3-10. The color code for a 100-ohm, $\frac{1}{2}$-watt 10% tolerance resistor is

    a. red, brown, black, gold

    b. brown, brown, brown, silver

    c. black, brown, black, silver

    d. brown, black, brown, silver

4

Circuit Laws

Several persons who developed the laws and rules for electrical phenomena were Ohm, Watt, Volta, and Ampere. Some of these men discovered how electricity works. Others identified rules with which to use electricity and how the various concepts of voltage, current, resistance, and power are interrelated. The most significant of these rules are called Ohm's law and Watt's law. These and one other called Kirchhoff's law are discussed in this chapter.

Persons involved in electrical or electronic work need rules with which to work. Each electrical circuit works in a certain way. Knowledge of the way the circuit works and how to identify each of the electrical concepts in the circuit is required if one is to be successful. There are very few "new" electrical circuits. Most of the "new" circuits are really modifications of another circuit. The rules identified by Ohm, Watt, and Kirchhoff apply to every electrical circuit. Quantities of voltage, current, resistance, and power are identified by applying the basic circuit laws developed by these men.

Once the electrical circuit is broken down into its component parts, it is better understood. The rules about to be presented are really very simple. They should be learned well by everyone planning to spend any amount of time working with electricity and electronics. Most circuit analysis is based upon an understanding of these laws. It is paramount that they be clearly understood. Not only should these laws be understood, but the technician must be able to apply the laws in practical work. The following sections cover the explanation of each of the laws and their applications.

BASIC ELECTRICAL CIRCUIT

Every electrical circuit that is capable of performing work has a minimum of three basic parts. These are illustrated in Figure 4-1. The

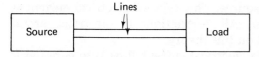

Figure 4-1. The basic electric circuit consists of a source, lines, and a load.

parts that are fundamental to each circuit are the source, the lines, and the load. The term source is used to represent a source of electrical energy. The source may be a battery, a dc or ac generator, or an electric power supply. The source provides the electromotive force and electrons required for the accomplishment of work. On the opposite side of the diagram is the load. The load represents a unit that is capable of converting electrical energy into a form of energy that does some work. The load in a simple circuit may be a light bulb. It may also be a complex computer or machine. In any case, the load is that part of an electrical circuit in which work is accomplished. If you remember that electrical power will produce some kind of work, the term "load" is easy to understand.

There is one other part of any electrical circuit. This part is the wires, or lines, that connect the load to the source. Under normal circumstances, two wires are required for this connection. These are the minimum requirements for all electrical circuits. In some cases two lines may not be obvious. There are many instances where the frame or case of the device is used in place of one of the lines. This is true in the automobile, where the body and frame are one conductor. One wire from the battery is connected to the body and frame. The other wire from the battery is connected to all electrical loads in the automobile.

Circuit Values

A review of the specific terms used to describe the electrical function of a circuit is in order at this time. These are just a few of the terms in common use. Learn them well and try to understand how each relates to electrical circuit action.

Current. Electrical current is the movement of charges in an electrical circuit. There are two schools of thought regarding electrical current. One uses electron theory as a standard. The electron is a negatively charged body. The assumption using this theory is that electron flow is from a highly concentrated field of electrons to a field in which there is a lack of electrons. Translating this into electrical terminology gives an electron flow from the negative connection on the source through one line to the load. The electron flow then moves through the load, through the second line, and back to the positive connection on the source. This type of electrical current is called electron current flow. All descriptions of electrical circuit action in this book are based on this theory.

A second way of treating electrical current flow is to consider it moving from the positive connection, through the load, and then to the negative source connection. This is called conventional current flow. The literature discussing electrical current flow uses either method. It really does not make a lot of difference in the performance of work. The main thing to keep in mind is that work is performed when an electrical current *moves* through a load. Very little work is accomplished when there is an absence of electrical current movement in the circuit.

Electron current is measured in amperes. The ampere describes the quantity of electrical charges moving past a given point in a circuit. There is a standard reference of time included in most electrical descriptions. This unit of time is the second. The quantity of electrons that flow past a given circuit point in 1 second is measured in units of the ampere.

The ampere is too large a unit to use in every description. There are, therefore, subunits. These are derived from the metric system of measure. The prefix *milli-* is used when describing 1/1000 of an ampere. One milliampere is equal to 0.001 ampere. The prefix *micro-* is used to describe electrical currents in the 1/1,000,000 range. One microampere is equal to 0.000,001 amperes. Another term is used to describe even smaller quantities. This is the *pico-*. The pico represents 1/1,000,000,000,000 of the whole unit. This in describing current is equal to 0.000,000,000,001 amperes. The prefixes milli-, micro-, and pico- are also used to describe other electrical units. They are fractional parts of the whole.

Voltage. The volt is the unit of electrical pressure or potential difference between two points in a circuit. Voltage exists in a circuit. Voltage is an electromotive force (EMF). This EMF forces electron movement in an electrical circuit. Each voltage source has a polarity at each moment of time. If the voltage source is dc, the polarity is constant. This is illustrated in Figure 4-2(a). The symbol for a battery is also given in this illustration. The load is represented with a resistance symbol. The + and − signs on the source indicate the polarity of the source.

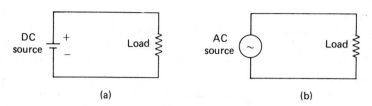

Figure 4–2. (a) DC circuit and its symbolic components; (b) AC circuit and its components.

Figure 4-2(b) also shows an ac source. The sine wave is used to represent the ac source. No polarity signs are shown because the polarity is constantly changing.

The quantity of EMF is measured in volts. The polarity of the voltage produces a difference of EMF between two points in an electrical circuit. This difference is measurable. The difference in EMF forces the electron movement in the circuit and gives the electrons a specific direction of travel. Electrons will move from a highly concentrated area to a less concentrated area in the circuit. The difference in EMF is also called a potential difference. Voltage has the potential to force electron flow in a circuit. Voltage, EMF, and potential difference are terms that are often used interchangeably.

Voltage drop. The term voltage drop is used to describe the difference in EMF between two points in a circuit. The movement of electrons through a resistance produces a voltage drop across that resistance.

Circuit ground or common. Every electrical circuit has a point of reference. This reference point is called by two names. One of these is *ground*. This term developed from the early days of electrical work. One wire from the source often was connected to the earth by means of a metal pipe in the ground. This point became a reference point. All circuit voltages were measured from "ground" to another point in the circuit.

Many of the circuits used today are not directly connected to ground. This led to the development of a secondary point of reference. This point is called *common*. The term common refers to a reference point in the electrical circuit. The common reference point may be directly connected to the earth. It may also be isolated from the earth. An example of this is found in the automobile. One side of the battery (usually the negative side) is called ground. It is directly wired to the body and frame. The rubber tires insulate the body from the contact with the earth. The ground, or common, on the automobile is a point of reference only for that vehicle. A second vehicle requires its own common. Both systems may be identical. They are, however, insulated from each other. Each has its own common reference point. One cannot measure from one vehicle to the other without first wiring both common systems to each other. Measurements of circuit values are often made by measuring the electron flow in the system. Electrons cannot move from the common on one vehicle to the other vehicle unless the two are wired together.

Common connections. Usually, the negative connection from the power source is connected to circuit common. This provides a circuit whose voltages are positive *with respect to circuit common.* This is not always the situation in current electronic devices. In some cases the common point is positive and all circuit voltages are negative with respect to common. In other situations a third reference point may be established. Then the circuit could have both positive and negative voltage values with respect to common. These three types of connections are illustrated in Figure 4-3. All three circuits have the

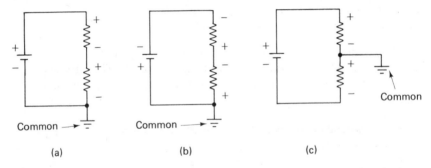

(a)   (b)   (c)

Figure 4–3. The common point of reference in any circuit may be anywhere in the circuit.

same power source and the same set of load components. The only difference in the circuits is the point at which the common connection is made.

Resistance. Resistance causes an opposition to the flow of electrons. All materials have some amount of resistance. As resistance values increase, the opposition to electrical current also increases. An increase in resistance in an electrical circuit will reduce the electrical current in the same circuit. Resistance is measured in units of ohms.

Voltage drop. Opposition to electric current in a circuit will establish a situation where there will be a different quantity of electrons that collect at the ends of a resistor. This different quantity is measurable. It is measured as a difference in electrical pressure, or EMF. The term *voltage drop* is used to describe this situation. Voltage drop is the difference in electrical pressure at two points in a circuit.

Short circuit. Under certain conditions the total resistance in an electrical circuit may be very low. This condition usually develops across the load. It is not desirable. A very low resistance will force large quantities of electric current flow. This may cause overheated wires and possibly a fire. If a battery is used as the source, it will exhaust the battery. This condition is known as a *short-circuit* condition. A short circuit means unusually low circuit resistance and very high current flow. It often results in damage to equipment.

In many cases a circuit protection device is added to the circuit. This is done in order to protect against damage in case of a short circuit. The protection device is either a fuse or a circuit breaker. These devices are designed to fail in cases of circuit overcurrent.

Open circuit. The open circuit is the opposite of the short circuit. In an open circuit there is an infinite amount of resistance. This is due to some component in the circuit failing. When some component fails, the electron path through the load is no longer working. Wires or components may break. The result is a separation of the current path. This is classified as a very high, or infinite, resistance. Electron flow in the circuit is reduced to zero and the circuit is no longer functional.

Circuit Rules

A common language has been established. This is the language of electrical theory, which is used to explain electrical concepts and their interaction. Now let's look at these rules.

Ohm's Law. Ohm's law explains the relationship of voltage, current, and resistance in a circuit. Ohm's law states that 1 ampere of current will flow through a resistance of 1 ohm when 1 volt of EMF is applied to the circuit. Certain letters are used to identify these quantities:

$$E = \text{EMF, in volts}$$

$$I = \text{current, in amperes}$$

$$R = \text{resistance, in ohms}$$

Ohm's law is shown as

$$E = I \times R$$

These quantities are mathematically related. It is possible to use Ohm's law to solve for each of the three quantities. The third quantity may be found as long as any two are known. The formulas are shown as

$$E = I \times R, \quad I = \frac{E}{R}, \quad R = \frac{E}{I}$$

Another way of showing this is to use either a wheel or a triangle with the formulas. This is illustrated in Figure 4-4. Here a triangle is

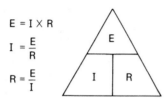

Figure 4-4. Ohm's law formulas. It is easier to remember the triangle arrangement rather than three formulas.

used. The voltage is at the top. The current and resistance may be in any order at the bottom. This triangle is used as an aid in remembering the relationship of E, I, and R. When using it, cover the section that identifies the unknown quantity. The equation is then established for finding this quantity.

Here are some examples of how this triangle is used to solve Ohm's law problems. Use the electrical circuit illustrated in Figure 4-5 as a reference.

Figure 4-5. A basic circuit used to determine circuit values with Ohm's law.

1. $E = 24$ V, $I = 2$ A; find R.

$$R = \frac{E}{I} = \frac{24}{2} = 12 \ \Omega$$

2. $I = 4$ A, $R = 8 \ \Omega$; find E.

$$E = I \times R = 4 \times 8 = 32 \text{ V}$$

3. $E = 100$ V, $R = 400 \ \Omega$; find I.

$$I = \frac{E}{R} = \frac{100}{400} = 0.25 \text{ A}$$

In each case, the letter designating the unknown is covered. The formula for finding this unknown is then shown in the exposed parts of the triangle.

PROBLEMS

Solve each of these problems using Ohm's law. Keep all answers in whole units of amperes, volts, and ohms.

1. $E = 24$ V, $R = 500,000$ Ω; find I.
2. $E = 12$ V, $R = 1,000,000$ Ω; find I.
3. $E = 130$ V, $I = 0.00012$ A; find R.
4. $E = 12$ V, $I = 0.00024$ A; find R.
5. $I = 0.0001$ A, $R = 50,000$ Ω; find E.
6. $R = 300,000$ Ω, I = 0.00032 A; find E.
7. Convert the answers to Problems 1 through 4 to units using scientific prefixes.

Watt's Law. Watt's law is directly related to power in the electrical circuit. This power may be in the form of useful work. For example, electric motors are often rated in units of horsepower. One horsepower is equal to 746 watts of electrical power.

$$1 \text{ hp} = 746 \text{ watts}$$

A one-half horsepower rated electric motor uses $\frac{1}{2}$ of 746, or 373 watts, of electric power.

Electric power relates to the time rate of doing work. The definition of electric wattage is 1 watt of power is used to perform work when 1 volt causes 1 ampere of current to flow for 1 second. The letter P is used to represent electrical power. Power is measured in units of the watt. The formula for electric power is

$$P = E \times I$$

The power formula relationships are similar to the Ohm's law relationship.

$$P = E \times I, \quad E = \frac{P}{I}, \quad I = \frac{P}{E}$$

These, too, may be inserted in a triangle similar to the one used when solving Ohm's law, as shown in Figure 4-6. The power is at the peak of this triangle. The voltage and the current may be in either of the

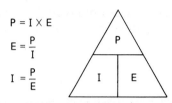

$P = I \times E$

$E = \dfrac{P}{I}$

$I = \dfrac{P}{E}$

Figure 4-6. Watt's law formulas. The triangle helps to remember the relationship between each value.

lower corners. To find the unknown quantity, cover that section of the triangle. The formula is as it appears in the remaining parts of the triangle.

Here are some examples of how this works. Use the power triangle to find each unknown.

1. $P = 40$ W, $E = 120$ V; find I.

$$I = \frac{P}{E} = \frac{40}{120} = 0.33 \text{ A}$$

2. $I = 3$ A, $P = 600$ W; find E.

$$E = \frac{P}{I} = \frac{600}{3} = 200 \text{ V}$$

3. $E = 12$ V, $I = 60$ A; find P.

$$P = E \times I = 12 \times 60 = 720 \text{ W}$$

This method uses the same procedure as does the Ohm's law triangle system.

PROBLEMS

1. $E = 120$ V, $I = 6$ A; find P.
2. $E = 600$ V, $I = 0.12$ A; find P.
3. $I = 0.0016$ A, $P = 32$ W; find E.
4. $I = 1.6$A, $P = 240$ W; find E.
5. $E = 240$ V, $P = 600$ W; find I.
6. $E = 600$ V, $P = 1.8$ W; find I.
7. Convert the answer in Problems 3 and 6 to units using the scientific prefixes.

Interrelationships of Ohm's and Watt's Laws. In mathematics, if a equals b, then b may be substituted for a. A may also be substituted for b. This is used to show interrelationships between the two laws. One example is

$$E = I \times R \quad \text{(Ohm's law)}$$
$$P = E \times I \quad \text{(Watt's law)}$$

$I \times R$ may be substituted for E in the Watt's law formula since they are equal.

$$P = E \times I$$
$$P = (I \times R) \times I$$

Combining like terms gives

$$P = I^2 R$$

All the interrelations between these two laws are shown in Figure 4-7. Note that this wheel is divided into four sections. Each section shows

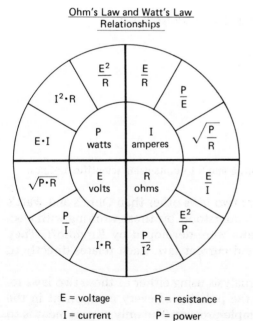

Ohm's Law and Watt's Law
Relationships

| E = voltage | R = resistance |
| I = current | P = power |

Figure 4-7. A wheel used to show the interaction between Ohm's and Watt's laws.

three formulas. Each formula may be used to solve for the quantity shown near the center of the circle. To use this, find the section that has the formula for the two known factors. Use this formula to solve for the unknown.

Examples of the use of this wheel are as follows:

1. $P = 100$ W, $I = 2$A; find R.

$$R = \frac{P}{I} = \frac{100}{4} = 25 \ \Omega$$

2. $P = 100$ W, $R = 25 \ \Omega$; find I.

$$I^2 = \frac{P}{R} = \frac{100}{25} = 4 = 2 \text{ A}$$

3. $R = 500 \ \Omega$, $I = 3$ A; find P.

$$P = I^2 R = 3^2 \times 500 = 9 \ \times 500 = 4,500 \text{ W, or } 4.5 \text{ kW}$$

PROBLEMS

Solve these problems using the Ohm's law-Watt's law wheel.

1. $E = 1,000$, $I = 10$; find P.
2. $E = 1,000$, $R = 10$; find P.
3. $E = 120$, $R = 10$; find P.
4. $P = 150$, $R = 96$; find I.
5. $I = 9$ A, $P = 20$ W; find R.
6. $I = 12$ A, $P = 60$ W; find R.
7. Convert the answers to problems 5 and 6 to units using scientific prefixes.

Kirchhoff's Laws. There are two laws other than Ohm's and Watt's laws that need to be clearly understood by those working with electrical circuitry. These two laws were developed by Kirchhoff. They are Kirchhoff's voltage law and current law. Each relates directly to circuit analysis.

A part of the process of analysis using either of these two laws requires the identification of the polarity of every component in the circuit. This is a relatively simple process. The only requirement is to follow the rules. An electrical circuit is shown in Figure 4-8(a). There are three components in this curcuit, R_1, R_2, and the dc source E_1. The system for assigning polarity is very simple. Start with the dc source. The longer line of the symbol used to represent a dc source is

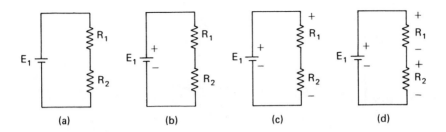

Figure 4-8. Development of polarities for each component in the circuit.

the positive terminal. This being the case, the shorter line is the negative terminal. Part (b) shows the labeling of the source polarity.

Now move from the source to the load side of the circuit. The top end of R_1 is directly wired to the positive terminal on the source. It is labeled with a + sign. The bottom end of R_2 is directly wired to the negative terminal on the source. It is labeled with a – sign. These additional signs are shown in part (c). A good general rule to follow is that the end of the component that is closest to a terminal is given the polarity of that terminal.

The final sign assignment is shown in part (d) of Figure 4-8. Each end of both resistors is given a polarity. The lower end of R_2 is closest to the negative terminal of the source, E_1. It is given a negative sign. The upper end of R_2 is farther away from the negative terminal. It is less negative than the other end of R_2. If it is less negative than the other end, it can be said to be positive. After all, it has fewer free electrons than does the negative end. Electrons move from high concentration to an area of less concentration. This is a direction of from negative to positive. If one end is negative, the other end of the same component receives the opposite sign. The same holds true for the other component in the circuit. R_1 is assigned a set of polarity signs using exactly the same set of rules. This is also illustrated in Figure 4-8(d). When completed, every component, including the source of power, has a set of polarity signs.

Kirchhoff's Voltage Law. The polarities of the components in any circuit help in understanding Kirchhoff's laws. His law relating to voltages is the first one explained. The law states that voltage drops in a closed loop add to equal zero. An example of this is shown in Figure 4-9. Part (a) shows one type of electrical circuit. This circuit has only one path for electron travel. It is *closed* to any other electron sources or forces. It is often called a *series circuit*. Kirchhoff called it a closed loop.

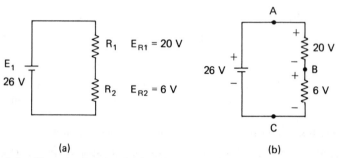

Figure 4-9. The voltage drops developed across each component in this circuit add to equal the source voltage.

Part (b) includes polarity assignments for all components. There is a difference in electrical potential across the terminals of each component in the circuit. This is called a *voltage drop*. These voltage drops may be added. Start at any point in the circuit. Travel either direction around the circuit until you return to the same point. Add each voltage drop and be sure to include the polarity of the drop as you do this. The first example starts at point A and travels in a clockwise direction. Use the letter E to represent each voltage drop across each component in the circuit. Subscripts of E_1, R_1, and R_2 are used to identify specific components.

$$-E_{E1} + E_{R1} + E_{R2} = 0$$

Substituting voltage values for these gives the formula

$$-26 + 20 + 6 = 0$$

Combine like terms.

$$0 = 0$$

Reversing the direction will give

$$+E_{E1} - E_{R2} - E_{R1} = 0$$
$$+ 26 - 6 - 20 = 0$$
$$0 = 0$$

If one starts at point B, the results are the same. Clockwise from point B,

$$+E_{R2} - E_{E1} + E_{R1} = 0$$
$$+6-26 + 20 = 0$$
$$0 = 0$$

Counterclockwise from point B,

$$-E_{R1} + E_{E1} - E_{R2} = 0$$
$$-20 + 26 - 6 = 0$$
$$0 = 0$$

The results are the same no matter where in the circuit one starts. Another way of remembering this law is to keep in mind that in a circuit containing one source and one or more loads the voltage drops across the loads will add to equal the voltage drop across the source. In the previous illustration, the drops add to equal 26 volts, also. This only works in circuits containing a single source.

There are other circuits that contain more than one source that are used in electric devices. When one finds a circuit of this type, the rules are still applicable. The first thing to do is to assign polarities to every component in the circuit. This is illustrated in Figure 4-10(a)

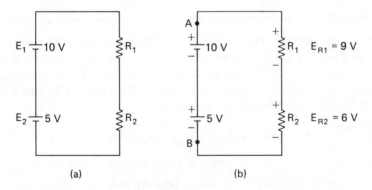

(a)   (b)

Figure 4-10. Two source voltages add when they have the same polarity.

and (b). The A circuit containing two or more sources requires some additional work before polarities are assigned to the loads. A determination as to the overall polarity of the total source must be made first. If both sources have the same polarity, as shown in parts (a) and (b), then simply add them. This gives a total EMF equal to the sum of the two source EMF's. The total EMF between points A and

B is 10 + 5, or 15 volts in this illustration. This is an application of Kirchhoff's voltage law. Applying this law to the whole circuit gives

$$+E_{R1} + E_{R2} - E_2 - E_1 = 0$$

$$+9 \text{ V} + 6 \text{ V} - 5 \text{ V} - 10 \text{ V} = 0$$

$$+15 - 15 = 0$$

$$0 = 0$$

If the source polarities are reversed, as illustrated in Figure 4-11(a), the rules still apply. Polarities are assigned to each compo-

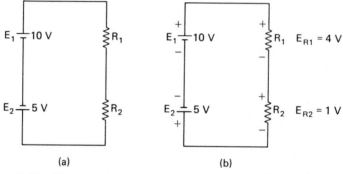

(a)   (b)

Figure 4-11.   Two source voltages subtract when they have opposing polarities.

nent in the circuit. The decision as to polarity for load components is based on the largest source EMF. In the case, E_1 is bigger than E_2. The polarity of E_1 is dominant in this circuit. These polarities are assigned to the load components. This is illistrated in Figure 4-11(b). Applying Kirchhoff's voltage law gives

$$+E_{R1} + E_{R2} + E_{E2} - E_{E1} = 0$$

$$+4 + 1 + 5 - 10 = 0$$

$$+10 - 10 = 0$$

$$0 = 0$$

This works in any multiple source circuit. In this circuit the voltage applied to the load is still the sum of the voltages from each source. Source E_2, being of reverse polarity, actually is subtracted from the voltage of source E_1. Since subtraction is actually addition of sums having opposite signs, the rules using Kirchhoff's laws are valid.

Practical solution of these circuits may require more information. How this is accomplished is illustrated in the next two chapters.

Kirchhoff's Current Law. The second Kirchhoff law relating to circuit analysis is his current law. This law relates to any junction of wiring in an electrical circuit. It states that the current entering a junction is equal to the current leaving the junction. This is illustrated in Figure 4-12. Current flow is from A to C in this circuit.

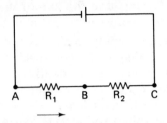

Figure 4-12. Kirchhoff's current law states that the current entering any junction is equal to the current leaving the same junction.

There is a junction of two resistors at point B. The current flowing through R_1 and entering junction B is equal to the current leaving junction B and flowing through R_2. This rule also is true for points A and C. Current from the source entering R_1 at point A is equal to the current leaving point A and entering R_1. This applies at point C also.

A circuit may contain more than one path in which electron current flows. Kirchhoff's current law also applies in this type of circuit. This type of circuit is shown in Figure 4-13. Both drawings show the

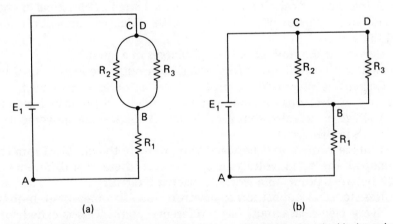

(a)   (b)

Figure 4-13. Two methods of showing a circuit. The preferred method is shown in part (a). Both of these circuits illustrate Kirchhoff's current law.

R_3 is also equal to the current at the positive terminal of the power source. This is expressed in the formula

$$I_{\text{source}} = I_{R1} = I_{R2} + I_{R3}$$
$$10\ A = 10\ A = 7\ A + 3\ A$$

same circuit. Part (b) is a more acceptable way of drawing the circuit. In this circuit there is only one path for current between points C and D. The current will divide in this part of the circuit. The amount of current in each of the paths depends upon the size of each resistance. Irregardless of this fact, the current divides at points B and C-D. The sum of the current at point B is equal to the sum of the currents through R_2 and R_3 in all cases. The sum of the currents of R_2 and

The rules expressed in this chapter apply to all types of electrical circuits. The technician needs to be able to identify specific electrical circuit configurations. Once the circuit type is identified, the specific rules are applied as the analytical process is completed. The result of this procedure provides a mathematical analysis of the electrical circuit. It also gives numerical values for voltage, current, resistance, and power in the circuit. Specifics for circuit analysis are given in the following three chapters.

SUMMARY

Circuit laws developed by Ohm, Watt, and Kirchhoff apply to all types of electrical circuits. These laws are used for the analysis of each electrical circuit.

A basic electrical circuit consists of a source of electrical energy, a load, and a pair of wires, or lines, with which to connect the load and the source.

Current is the flow of electrical charges in a circuit.

Electron current flows fom negative through the load to positive.

Current is measured in units of the ampere. It represents the quantity of electrons that move past a given circuit point in 1 second.

Voltage is an electromotive force. It causes the movement of electrons in the circuit.

Voltage is measured between two points in the circuit. From this developed the term voltage drop, because there is a difference in EMF between points in a working electrical circuit.

Resistance is opposition to electron flow. It is measured in ohms.

Every electrical circuit has a reference point. This is called common or ground. Circuit voltage measurements are usually made using common as a reference point.

Voltages may be either positive or negative with respect to circuit common. This depends upon the wiring of the source to the load.

An electrical short circuit occurs in an electrical circuit when there is an undesirable low resistance path in the circuit. This produces an unusually large current flow in the circuit and tends to produce heat. This may cause a fire or destruction of the remaining circuit components.

An open circuit is the interruption of circuit current. This may be caused by the failure of a component. It may be the result of turning the circuit switch to the "off" position.

Ohm's law states that 1 ampere of current will flow through 1 ohm of resistance when 1 volt of EMF is applied to the circuit. $E = I \times R$.

Watt's law relates to electric power. One watt of power is used when 1 volt causes 1 ampere of current to flow for 1 second. Power is the time rate of performing work. $P = E \times I$.

Ohm's law and Watt's law are interrelated. Equal terms may be substituted from one formula to the other.

All electrical circuits have a polarity. The polarity is developed by the power source.

Kirchhoff's voltage law states that all the voltage drops developed in a closed-loop electric circuit add to equal zero.

Kirchhoff's current law states that the current entering an electrical junction is equal to the current leaving the same junction.

FORMULAS

Ohm's Law:

$$E = I \times R$$

Watt's law-

$$P = I \times E$$

Kirchhoff's voltage Law:

$$E_{\text{source}} + E_{R1} + E_{R2}, \text{etc.} = 0$$

Kirchhoff's current law:

I entering junction 1 = I leaving junction 1

QUESTIONS

4-1. The basic electric circuit consists of

    a. load, lines, source

    b. current, wires, source

      c. voltage, current, lines

      d. source, load, voltage

4-2. Electron current flow is

      a. from negative to positive in the source

      b. from negative to positive in the load

      c. from positive to negative in the load

      d. dependent upon the polarity of the source

4-3. Ohm's law is stated as

      a. $I = E \times R$

      b. $R = E \times I$

      c. $E = I \times R$

      d. all of the above

4-4. Watt's law is stated as

      a. $P = E \times I$

      b. $I = P \times E$

      c. $E = I \times P$

      d. $P = R^2 \times I$

4-5. When all voltage drops add up to zero this statement is related to

      a. Kirchhoff's voltage law

      b. Kirchhoff's current law

      c. Ohm's law

      d. Watt's law

4-6. Electrical current entering a junction is ____ to current leaving a junction.

      a. equal to

      b. more than

      c. less than

      d. has no effect on

4-7. A closed-loop circuit has three resistive elements. These elements develop voltage drops of 10, 15, and 25 volts when current flows. The source voltage is ____ volts.

      a. 25

      b. 15

      c. 10

      d. 50

4-8. Three branch circuits are connected to one main circuit. The main circuit has a current flow of 24 amperes. One branch has a current of 6 amperes and the second branch has a current of 10 amperes. The third branch current is

a. 4 A

b. 6 A

c. 8 A

d. 10 A

4-9. Kirchhoff's circuit laws apply

a. to ac circuits

b. to dc circuits

c. during rising current periods

d. to all electrical circuits

4-10. If 240 volts is applied to a load with a resistance of 12 ohms, the current flow in the load is

a. 0.05 A

b. 1.5 A

c. 10 A

d. 20 A

5

Series Circuits

There are two basic circuits found in electrical work. These are the series circuit and the parallel circuit. Any other circuit that is used is a modification of one of these two circuits. Each one of these has its own set of rules. The rules apply to current, resistance, voltage drops, and power in the circuit. Circuit anlaysis is based upon recognition of the type of circuit and application of the related rules for that circuit. The circuit explained in this chapter is the series circuit. It is illustrated in Figure 5-1. It consists of a source, one or more loads, and connecting lines.

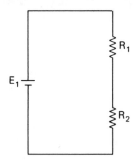

Figure 5-1. The basic series circuit with one source and two loads.

IDENTIFICATION OF THE SERIES CIRCUIT

The series circuit is easy to identify. The major means of identification in this circuit is the number of paths in which current can flow. The series circuit has only one path. It may be drawn as illustrated in Figure 5-1. Often this circuit is simplified. The power source is not illustrated. A number representing the voltage value is shown instead. This number is placed at one end of the components making up the circuit. Usually, this is the positive end of the circuit. This is shown in Figure 5-2. When this type of circuit is shown, the connections to the power source are assumed.

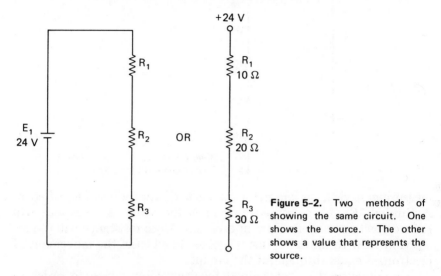

Figure 5-2. Two methods of showing the same circuit. One shows the source. The other shows a value that represents the source.

RESISTANCE IN THE SERIES CIRCUIT

Resistance in the series circuit depends upon the number of resistive loads in the circuit and the resistance value of each of these loads. Resistance values are added in the series circuit. Total circuit resistance is equal to the sum of each resistance in the circuit.

$$R_{\text{Total}} = R_1 + R_2 + R_3, \text{etc.}$$

As additional resistors are wired into the circuit, the total resistance increases as a result of the addition of their ohmic values in the circuit.

CURRENT IN THE SERIES CIRCUIT

Current in a series circuit depends upon two factors. These are the total circuit resistance and the source voltage. There is but one path for current flow in the series circuit. This being the case, the circuit current is equal throughout the circuit. Figure 5-3 illustrates a series circuit with three load resistances. If one inserts a current-

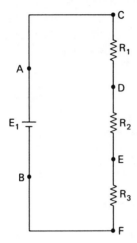

Figure 5–3. A series circuit used as an example in the text

measuring device, or ammeter, at each of the points identified as *A* through *F*, the measured current is the same. A resistance will reduce electron current flow in a circuit. More resistance will reduce electron flow to an even greater degree. In all cases the movement of electrons is equal throughout the circuit.

Circuit current is determined by using Ohm's law. In order to find the *R* for this formula, total resistance is determined by adding each resistance vlaue.

$$R_T = R_1 + R_2 + R_3 = 10 + 20 + 30 = 60 \ \Omega$$

The total resistance value is inserted into the formula $I = E/R$. Substituting known values for the letters given,

$$I = \frac{12}{60} = 0.2 \ \text{A}$$

Each resistance has a current flow of 0.2 amps. The ohmic value of the individual resistor does not change total current flow.

VOLTAGE DROPS

Voltage drop is defined as a difference in electrical pressure developed across a resistance as current flows in a circuit. In the series circuit the voltage drop developed across each resistive component is always less than the source voltage. The illustrations in Figures 5-4 and 5-5 will aid in understanding this situation.

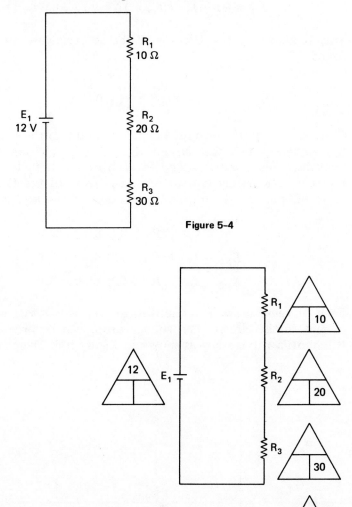

Figure 5-4

Figure 5-5. Add Ohm's law triangles and given information in order to solve for unknown values.

Ohm's law triangles are placed next to each component in the circuit. All the information known about the circuit is placed in the appropriate section of each triangle. At this moment there does not appear to be enough information. This is not true, however.

Total current can be found by adding the resistance values and applying Ohm's law.

$$R_T = R_1 + R_2 + R_3 = 10 + 20 + 30 = 60 \ \Omega$$

This is added to the triangle at the source. Now total current is found.

$$I_T = \frac{E_T}{R_T} = \frac{12}{60} = 0.2 \ \text{A}$$

Current is equal throughout the series circuit. If the total current is 0.2 amperes, then the current through each resistance is also 0.2 amperes. This value is added to each of the triangles next to the resistances. Knowing two of the three parts of the Ohm's law triangle will produce the third part. It is done in this manner:

$$E_{R1} = I_{R1} \times R_1 = 0.2 \times 10 = 2 \ \text{V}$$

$$E_{R2} = I_{R2} \times R_2 = 0.2 \times 20 = 4 \ \text{V}$$

$$E_{R3} = I_{R3} \times R_3 = 0.2 \times 30 = 6 \ \text{V}$$

Each of these answers is the voltage drop across that specific component in the circuit. The voltage drops can be proved by use of Kirchhoff's voltage law as shown in Figure 5-6. The series circuit is

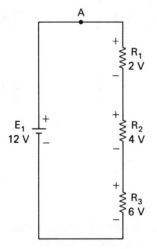

Figure 5-6. Use Kirchhoff's voltage law to solve for unknown values.

a closed loop. Start at point *A* and move in a clockwise direction around the loop. Add all the voltage drops.

$$+2 + 4 + 6 - 12 = 0$$

$$+12 - 12 = 0$$

$$0 = 0$$

ANALYSIS OF THE SERIES CIRCUIT

There are many times when the electrical technician needs to know how to analyze a circuit. This analysis provides information relative to current flow, resistance, power and voltage drops for every component in the circuit. Different methods are used to accomplish this analysis. The method illustrated in this section is one of the basic methods in use today.

The circuit illustrated in Figure 5-7 is used as the example. It

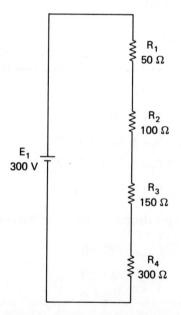

Figure 5–7

is a series circuit that contains four resistances and one source. Use a separate sheet of paper when solving this problem. Follow each step as illustrated.

Step 1. Place an Ohm's law triangle next to each component and the source. Fill in all known information. This is the source voltage and the value of each resistor.

Step 2. Review the circuit information to determine what information is most likely to be of assistance in the analysis process. In the case of this series circuit, finding total current will provide current flow through each of the resistors. This leads to power and voltage drop information.

Step 3. Find total resistance.

$$R_T = R_1 + R_2 + R_3 + R_4 = 50 + 100 + 150 + 300$$
$$= 600 \ \Omega$$

Step 4. Find total current.

$$I_T = \frac{E_T}{R_T} = \frac{300}{600} = 0.5 \text{ A}$$

Step 5. Find voltage drops across each resistance. The value of current identified in step 4 is added to each triangle since the current is equal throughout the circuit.

$$E_{R1} = I_{R1} \times R_1 = 0.5 \times 50 = 25 \text{ V}$$
$$E_{R2} = I_{R2} \times R_2 = 0.5 \times 100 = 50 \text{ V}$$
$$E_{R3} = I_{R1} \times R_3 = 0.5 \times 150 = 75 \text{ V}$$
$$E_{R4} = I_{R4} \times R_4 = 0.5 \times 300 = 150 \text{ V}$$

Step 6. Verify the voltage drops using Kirchhoff's law.

$$E_{R1} + E_{R2} + E_{R3} + E_{R4} + E = 0$$
$$25 + 50 + 75 + 150 - 300 = 0$$
$$0 = 0$$

Step 7. Find the power for each resistance value and the source. There is no order to this. Use the power formula and/or triangle as Ohm's law is used.

$$P_T = E_T \times I_T = 300 \text{ V} \times 0.5 = 150 \text{ W}$$
$$P_{R1} = E_{R1} \times I_{R1} = 25 \times 0.5 = 12.5 \text{ W}$$

or

$$P_{R1} = I_{R1}^2 \times R_1 = 0.5^2 \times 50 = 12.5 \text{ W}$$

$$P_{R2} = E_{R2} \times I_{R2} = 50 \times 0.5 = 25 \text{ W}$$

$$P_{R3} = E_{R3} \times I_{R3} = 75 \times 0.5 = 37.5 \text{ W}$$

$$P_{R4} = E_{R4} \times I_{R4} = 150 \times 0.5 = 75 \text{ W}$$

Step 8. Verify the power. The total power in a series circuit is equal to the sum of the power in each component.

$$P_T = P_{R1} + P_{R2} + P_{R3} + P_{R4}$$
$$= 12.5 + 25 + 37.5 + 75$$
$$= 150 \text{ W}$$

A second circuit is offered in order to show how to solve for circuit values when not all circuit values are given. Figure 5-8 is the

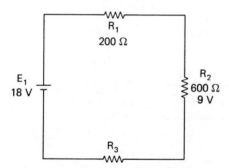

R_1
200 Ω

E_1
18 V

R_2
600 Ω
9 V

R_3

Figure 5-8

circuit being analyzed. The following steps are given in a way to analyze this circuit. The voltage drops, current, resistance, and power for each component is to be found.

Step 1. Place an Ohm's law and a Watt's law triangle next to each component and the source. Fill in all known information as given on the diagram.

Step 2. Review the circuit in order to plan the approach to the solution of this problem. In this case there is only one component that has sufficient information to use as a starting point. This is R_2. Both the voltage drop and the resistance are given. Finding the current through R_2 will provide a current value for the whole circuit. This information will then aid in solving for the other unknown values.

Step 3.   Find the current through R_2.

$$I_{R2} = \frac{E_{R2}}{R_2} = \frac{9}{600} = 0.015 \text{ A}$$

Place this answer in all other triangles. At this point any component's values may be found with the exception of R_3. Fill in the balance of the unknowns for E_1, R_1, and R_2 in any order.

$$P_{R1} = I_{R1}^2 \times R_1 = 0.000225 \times 200 = 0.045 \text{ W}$$

$$E_{R1} = I_{R1} \times R_1 = 0.015 \times 200 = 3 \text{ V}$$

$$P_{R2} = E_{R2} \times I_{R2} = 9 \times 0.015 = 0.135 \text{ W}$$

$$R_T = \frac{E_T}{I_T} = \frac{18}{0.015} = 1200 \ \Omega$$

$$P_T = E_T \times I_T = 18 \times 0.015 = 0.27 \text{ W}$$

Step 4.   Use Kirchhoff's law to find the voltage drop across R_3.

$$E_{R1} + E_{R2} + E_{R3} + E = 0$$

$$3 + 9 + E_{R3} - 18 = 0$$

$$E_{R3} = 18 - 3 - 9$$

$$E_{R3} = 6 \text{ V}$$

Step 5.   Use Ohm's law to find the value of R_3.

$$R_3 = \frac{E_{R3}}{I_{R3}} = \frac{6}{0.015} = 400 \ \Omega$$

Step 6.   Use Watt's law to find the power of R_3.

$$P_{R3} = E_{R3} \times I_{R3} = 6 \times 0.015 = 0.09 \text{ W}$$

This approach is used to solve for circuit values in any series circuit. The approach is modified as required. This depends upon the specific information given for the circuit.

OPEN CIRCUITS AND SHORT CIRCUITS

In any working circuit there may be breakdowns. These are usually classified as either opens or shorts. When either one of these occurs, the values in the circuit change. The circuits shown in Figure 5-9 are typical of many found in electronic systems. R_2 could be a

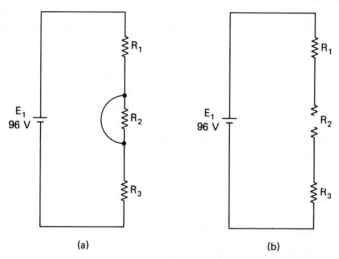

(a)   (b)

Figure 5-9. Short circuit and open-circuit situations alter typical circuit values.

device other than a resistor. This device exhibits resistive qualities. A very common device of this type is the transistor. For the sake of discussion, R_2 will be identified as part of a transistor.

The values of each of the three resistors is 500 ohms. The source voltage is 96 volts. Using Ohm's law, the current in the circuit is 64 milliamperes. The voltage drop across each resistance is 32 volts. The power in each unit is 2.05 watts. Now see what happens to the circuit values if the transistor were to fail and the internal parts short together. This is shown in part (a). The resistance of R_2 drops to zero ohms. Total circuit resistance decreases to 1,000 ohms. Circuit current increases from 64 to 96 milliamperes. This is a 50% increase in current in the circuit. The power in R_1 increases from 2.05 to 4.6 watts due to the increased current flow. The voltage drop across R_2 goes to zero. The voltage drops across R_1 and R_3 increase to 48 volts each. This increase in power, voltage, and current may cause another component failure in the circuit. The increases have probably produced circuit values that are higher than the circuit design factors. This will produce a failure in the other components.

Assume that the transistor of Figure 5-9 represented as R_2 failed by "burning out." In other words, it created an open-circuit condition as it failed. The resistance of the transistor has increased from its original 500-ohm value. Its new resistance value is several million ohms. The resistance could be identified as an infinitely high value, too high to measure with standard measuring equipment.

For the sake of discussion, the resistance value of the transistor R_2 is 500 megaohms. Total circuit resistance has gone from 1,500 to 500,001,000 ohms. Using Ohm's law, the circuit current is about 2 microamperes. The voltage drop developed across each component has changed. In the original working circuit, each resistor had a voltage drop of 32 volts. In the open circuit the drop across R_3 is about 95 volts. This amount of voltage is almost equal to the source voltage. The voltage drops that develop across R_1 and R_3 are close to zero. This is the situation in most open-circuit conditions. The normal voltage drop across each component changes. Almost all the applied source voltage appears across the open component in the circuit.

FORMULAS

Resistance: $R_T = R_1 + R_2 + R_3$, etc.

Current: $I_T = I_{R1} = I_{R2} = I_{R3}$, etc.

Voltage: $E_T = E_{R1} + E_{R2} + E_{R3}$, etc.

Power: $P_T = P_{R1} + P_{R2} + P_{R3}$, etc.

SUMMARY

One of the two basic circuits used in electrical work is the series circuit. The series circuit is easy to identify because it has only one path for electron flow. It may consist of one or more components and a power source. The series circuit has some characteristics that are only found in this circuit configuration.

Resistance values in the series circuit are added. Total circuit resistance is equal to the sum of each of the individual resistive values in the circuit.

Current in this circuit is the same value at each and every point in the circuit. Total circuit current is dependent upon the total resistance and the applied voltage. This is determined by the use of Ohm's law.

Voltage drop is the term used to identify a difference in electrical potential found in a circuit. A voltage drop is produced when current flows through a resistance. The sum of each voltage drop in the series circuit will always equal the applied voltage.

The terms 'open' and 'short' often confuse people. An open circuit is one that has a very high resistance value. This is often the result of a break in the current path. The open circuit is said to have infinite resistance. A short circuit is one that has a very low resistance value. It usually is in the order of less than one ohm of resistance. Current flow in the open circuit is zero. Current flow in the short circuit is very high.

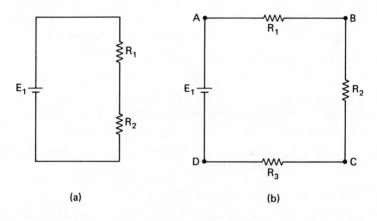

(a)   (b)

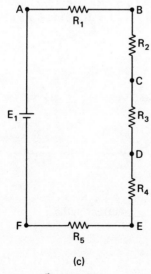

(c)

Figure 5–10

PROBLEMS

Solve each problem for the unknowns. Use Figure 5-10(a) for these problems.

5-1. $R_1 = 56\ \Omega, R_2 = 47\ \Omega, E = 2.0$ V; find I_T and E_{R1}.

5-2. $R_1 = 3.3$ kΩ, $R_2 = 2.2$ kΩ, $I_T = 55$ mA; find E.

5-3. $R_1 = 10$ kΩ, $R_2 = 500\ \Omega, E = 15$ V; find E_{R1}, E_{R2}, I_T, and P_T.

5-4. $R_1 = 1$ kΩ, $R_2 = 100\ \Omega, I_T = 10$ mA; find E_T and P_T.

Use Figure 5-10(b) for these problems:

5-5. $E = 30$ V, $I_T = 150$ mA. All resistors have equal value. Find R_1.

5-6. $R_1 = 470\ \Omega, R_2 = R_3, E = 12$ V; find E_{R2}.

5-7. $R_1 = 470\ \Omega, R_2 = 330\ \Omega, R_3 = 220\ \Omega, E = 51$ V; find I_T and P_{R1}, P_{R2}, P_{R3}.

5-8. $R_1 = 150\ \Omega, R_2 = 1{,}300\ \Omega, R_3 = 220\ \Omega$, and $E = 50$ V; find I_T, E_{R1}, E_{R3}, P_{R2}, and the voltage drop developed between points A and C.

5-9. $R_1 = 30\ \Omega, R_2 = 10\ \Omega, I_T = 0.1$ A, $E_{R3} = 2$ V; find E and R_3.

5-10. $R_1 = 5$ kΩ, $R_2 = 6$ kΩ, $R_3 = 4$ kΩ, and $I_T = 50$ mA; find the voltage at point D with respect to point B.

Use Figure 5-10(c) for these problems

5-11. All resistor values are 500 ohms. Source voltage is 50 volts. Find current, voltage drop, and power values for each component in the circuit.

5-12. All resistor values except R_2 are 1,000 ohms. Current through R_3 is 100 mA. Power in R_2 is 20 watts. Find R_2 and source voltage.

QUESTIONS

5-1. Resistance in a series circuit is ____ the sum of each resistance in the circuit.

    a. less than

    b. more than

    c. equal to

    d. not affected by

5-2. A series circuit has ____ current paths.

    a. one

    b. two

    c. multiple

    d. variable

5-3. Current in a series circuit is _____ source current.

    a. less than

    b. more than

    c. equal to

    d. not affected by

5-4. Current flow in a series circuit develops a _____ across each component.

    a. voltage drop

    b. power drop

    c. current drop

    d. equalizing factor

5-5. An open circuit has a _____ resistance at the open point.

    a. low

    b. high

    c. open

    d. variable

5-6. A short circuit has a _____ resistance at the shorting point.

    a. low

    b. high

    c. open

    d. variable

5-7. Increasing source voltage in a series circuit will

    a. increase current flow

    b. increase resistance

    c. decrease power dissipation

    d. not affect the circuit

5-8. Increasing load resistance in the series circuit will

    a. increase current flow

    b. decrease current flow

    c. increase power dissipation

    d. decrease individual voltage drops

5-9. Increasing source voltage for a series circuit will

    a. increase power dissipation

    b. decrease individual voltage drops

    c. decrease current flow

    d. increase circuit resistance

5-10. Three 100-ohm resistances are connected in series. A 200-volt source is applied to the resistors. The voltage drop developed across each resistor is about

    a. 33 volts

    b. 50 volts

    c. 67 volts

    d. 200 volts

6

Parallel Circuits

The second major circuit configuration is called the parallel circuit. This circuit differs from the series circuit in that it always has multiple paths for current flow. The series circuit has only one current flow path. The parallel circuit consists of a source, two or more loads, and connecting lines.

The parallel circuit is probably used more in electrical work than any other type of circuit. It is illustrated in Figure 6-1. The typical

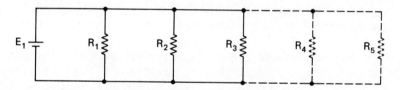

Figure 6-1. A parallel-wired circuit may have two or more branches.

parallel circuit has one source and two or more loads. The loads in this figure are represented by the resistances R_1 through R_5. The dashed lines are shown in order to represent any additional number of loads in the circuit.

Each resistive loads is considered to be directly connected to the power source. Drawing the circuit in a manner that showed this could make a very confusing drawing. This is illustrated in Figure 6-2. Figure 6-1 is much easier to understand. It also makes circuit analysis easier. Drawings for parallel circuits use the preferred method shown in Figure 6-1.

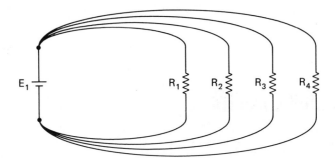

Figure 6-2. Actual wiring of a parallel circuit may be as shown. The preferred method of drawing the circuit is shown in Figure 6-1.

VOLTAGE IN THE PARALLEL CIRCUIT

One of the easiest parts to understand in the parallel circuit is the voltage drops across each of the components. This is simple because they are all equal. Each load is wired directly across the source. The formula for voltage drops in the parallel circuit is

$$E_{\text{source}} = E_{R1} = E_{R2} = E_{R3} \text{, etc.}$$

One of the most common parallel circuit applications is found in the home. Almost all home wiring consists of parallel circuits. Several "branches" are wired into the distribution panel. Each has the same voltage. It is 120 volts as supplied by the power company. All home appliances except perhaps a stove or air conditioner use the 120-volt source.

A second example of this type of circuit is found in the automobile. All the loads in the car require a 12-volt source. Each load is connected to the battery terminals or to a wire that is connected to the battery terminals.

CURRENT IN THE PARALLEL CIRCUIT

Currents in the parallel circuit are also reasonably simple to understand. Each section of a parallel circuit is called a *branch circuit*. The current through one branch circuit adds to the current of every other branch circuit. This provides a total current flow. The formula for this is

$$I_{\text{total}} = I_{R1} + I_{R2} + I_{R3} \text{, etc.}$$

The determination of currents in a parallel circuit is done in the following manner:

1. Draw or sketch the diagram of the circuit.
2. Label all components (E_1, R_1, R_2, etc.).
3. Add all known information for each component.
4. Determine individual unknown values.
5. Add currents.

Use Figure 6-3 as a guide, and let

$$E_1 = 25 \text{ V}$$
$$R_1 = 500 \text{ }\Omega$$
$$R_2 = 1,000 \text{ }\Omega$$

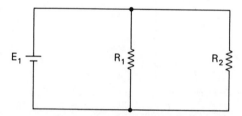

Figure 6-3. A parallel circuit for the example given in the text.

It is often less confusing if one draws an Ohm's law triangle next to each component in the circuit. This is shown in Figure 6-4(a). Then fill in the proper section of the triangle with all known information as shown in part (b). The information provided for each component is not sufficient to solve for currents. More information is required. Apply the rule for voltage in the parallel circuit at this time. Voltage is equal throughout the circuit. If this is true, then the voltage across R_1 and R_2 is the same as the voltage across the source. Add the 25-volt values to the triangles for R_1 and R_2.

Two of the three factors of Ohm's law are now known for R_1 and R_2. The third is found by using Ohm's law. Use a formula that includes the subscript R_1 for each value. This will make the process less confusing.

$$I_{R1} = \frac{E_{R1}}{R_1}$$

$$= \frac{25}{500}$$

$$= 0.05 \text{ A}$$

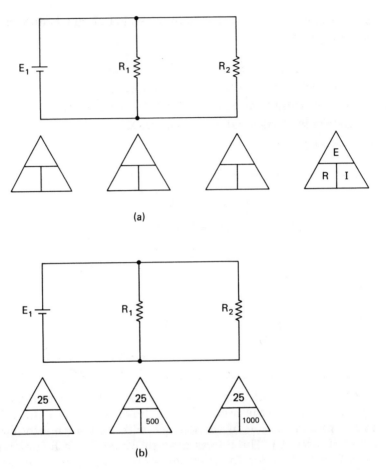

(a)

(b)

Figure 6–4. (a) The solution for circuit values is made easier by use of Ohm's law triangles; (b) Add all information given in the text before starting to solve for circuit values.

Repeat the process to find the current through R_2.

$$I_{R2} = \frac{E_{R2}}{R_2}$$

$$= \frac{25}{1,000}$$

$$= 0.025 \text{ A}$$

Once the current through each branch is found, the total current in the circuit is determined.

$$I_{\text{total}} = I_{R1} + I_{R2}$$
$$= 0.05 + 0.025$$
$$= 0.075 \text{ A}$$

This value may be inserted in the triangle used at the source. The addition of the total current to this triangle allows the determination of the total resistance in this circuit.

$$R_{\text{total}} = \frac{E_{\text{total}}}{I_{\text{total}}}$$
$$= \frac{25}{0.075}$$
$$= 333.3 \ \Omega$$

Use the schematic drawing shown in Figure 6-4. Change the values to

$$R_1 = 24 \ \Omega, \ R_2 = 48 \ \Omega, \ E_1 = 6 \text{ V}$$

Find I_{R1}, I_{R2}, and I_T.

$$I_{R1} = \frac{E_{R1}}{R_1} = \frac{6}{24} = 0.25 \text{ A}$$

$$I_{R2} = \frac{E_{R2}}{R_2} = \frac{6}{48} = 0.125 \text{ A}$$

$$I_T = I_{R1} + I_{R2} = 0.25 + 0.125 = 0.375 \text{ A}$$

TOTAL RESISTANCE IN THE PARALLEL CIRCUIT

The source in any circuit reacts to all the loads. It has no way to distinguish the currents or resistive values of individual loads. The source is said to look at an effective resistance value. This effective resistance is developed by the characteristics of each individual load. The effective resistance represents the total of each load.

A general rule to keep in mind when considering the relationships of voltage, current, and resistance in any circuit is *if the voltage is held to a constant value, the current rises as the circuit resistance drops.*

Conversly, if the voltage is constant and the resistance rises, the current will drop to a lower value. This is proven by Ohm's law.

This rule may be applied to a parallel circuit with the same results. Adding more loads to the total circuit increases the total current flow. This indicates that a reduction in total resistance has occurred. This is illustrated in an example using Figure 6-5.

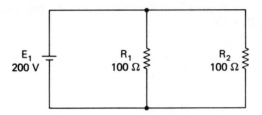

Figure 6-5

First determine the currents through each branch in this circuit.

$$I_{R1} = \frac{E_{R1}}{R_1} = \frac{200}{100} = 2 \text{ A}$$

$$I_{R2} = \frac{E_{R2}}{R_2} = \frac{200}{100} = 2 \text{ A}$$

Next determine the total current.

$$I_T = I_{R1} + I_{R2} = 2 + 2 = 4 \text{ A}$$

The next step is to determine total resistance.

$$R_T = \frac{E_T}{I_T} = \frac{200}{4} = 50 \text{ } \Omega$$

The total effective resistance is less than the value of each individual resistor in the parallel circuit. This is always true.

The example shown uses two equal resistance values as loads. When all the loads have the same value, total resistance may be determined by dividing the number of equal resistors into the ohmic value of one of the resistors. This is true regardless of the quantity of equal resistors used in the circuit. The formula for this is

$$R_T = \frac{R_1}{R_n}$$

where R_1 = one resistor value

R_n = quantity of equal resistors

If the resistor values in a parallel circuit are not equal, a different way of determining total resistance is used. The procedure for this is to add the sum of the reciprocals of each of the resistors. There are two ways of doing this. Use Figure 6-6 as the circuit for this solution.

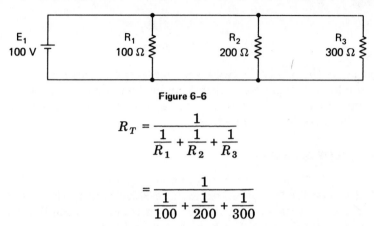

Figure 6-6

$$R_T = \cfrac{1}{\cfrac{1}{R_1} + \cfrac{1}{R_2} + \cfrac{1}{R_3}}$$

$$= \cfrac{1}{\cfrac{1}{100} + \cfrac{1}{200} + \cfrac{1}{300}}$$

This fraction cannot be added until the denominators of each fraction are equal.

$$R_T = \cfrac{1}{\cfrac{6}{600} + \cfrac{3}{600} + \cfrac{2}{600}}$$

$$= \cfrac{1}{\cfrac{11}{600}}$$

$$= \frac{600}{11} \times \frac{1}{1} = 54.5 \ \Omega$$

A simpler way, if one has a calculator, is to find the reciprocal of each number and add these. Then find the reciprocal of the total. This will give the correct answer.

$$R_T = \cfrac{1}{\cfrac{1}{100} + \cfrac{1}{200} + \cfrac{1}{300}}$$

$$= \frac{1}{0.01 + 0.005 + 0.003}$$

$$= \frac{1}{0.0183}$$

$$= 54.6 \ \Omega$$

The answer is slightly different owing to the rounding off of the numbers.

When two resistors of unequal value are connected in parallel, the *product over the sum* formula may be used. This formula is

$$R_T = \frac{R_1 \times R_2}{R_1 + R_2}$$

In each case the total effective resistance in the circuit is always lower than the lowest single resistance value in the circuit. If the answers to problems do not agree with this statement, the answer is incorrect. Check the work to find the error.

CURRENT DIVIDERS

If the resistance values for each resistive load in the circuit are known and the total current in the circuit is known, each branch current may be determined. The branches are said to be *current dividers*. Kirchhoff's current law applies in this situation. One may use Ohm's law to determine exact current values. One may also use a ratio method. Both are shown in the following examples. Figure 6-7 is the circuit used as the model.

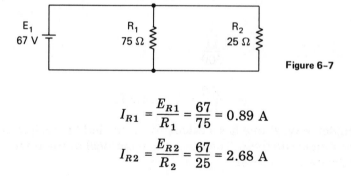

Figure 6-7

$$I_{R1} = \frac{E_{R1}}{R_1} = \frac{67}{75} = 0.89 \text{ A}$$

$$I_{R2} = \frac{E_{R2}}{R_2} = \frac{67}{25} = 2.68 \text{ A}$$

The total current in this circuit is

$$I_T = I_{R1} + I_{R2} = 0.89 + 2.68 = 3.57 \text{ A}$$

Solving for individual currents shows that the greatest amount of current flows through the smaller of the two resistive loads. It is possible to establish a ratio of load resistances and then use this ratio to determine the current through each resistance.

$$R_1 = 75, \quad R_2 = 25$$

Therefore, there are four parts to the two resistances. R_1 has three quarters of the total and R_2 has one quarter of the total.

$$I_T = 3.57 \times \tfrac{3}{4} = 2.69 \text{ A}$$

$$I_T = 3.57 \times \tfrac{1}{4} = 0.89 \text{ A}$$

This is an excellent way to determine the branch current values in a parallel circuit.

POWER IN THE PARALLEL CIRCUIT

The power formula states that power is the product of voltage times current, or $P = E \times I$. If the voltage in a parallel circuit is held to a constant value, increasing the current will also increase the power in that circuit. Total power in a parallel circuit is the sum of the power in each branch of that circuit. Use Figure 6-8 as a reference for this example.

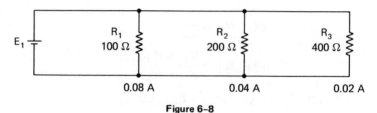

Figure 6-8

$$P_{R1} = (I_{R1})^2 R_1 = (0.08)^2 \times 100 = 0.0064 \times 100 = 0.64 \text{ W}$$

$$P_{R2} = (I_{R2})^2 R_2 = (0.04)^2 \times 200 = 0.0016 \times 200 = 0.32 \text{ W}$$

$$P_{R3} = (I_{R3})^2 R_3 = (0.02)^2 \times 400 = 0.0004 \times 400 = 0.16 \text{ W}$$

$$P_T = P_{R1} + P_{R2} + P_{R3} = 0.64 + 0.32 + 0.16 = 1.1 \text{ W}$$

If one wished to find the source voltage for this example, an application of Ohm's law is all that is required. Any of the loads may be used for this. This is the case because all the loads are parallel wired to the source. All voltages are equal.

$$E_{R1} = I_{R1} \times R_1 = 0.08 \times 100 = 8 \text{ V}$$

SUMMARY

The two basic electrical circuits are called parallel and series.
In the parallel circuit

1. Voltage is equal across all loads.
2. Total current is the sum of each branch current.
3. Total resistance is the sum of the reciprocals of each branch resistance.
4. Total power is the sum of each branch power.
5. Current divides in inverse proportion to load resistance.

FORMULAS

Voltage: $E = E_{R1} = E_{R2} = E_{R3}$, etc.

Current: $I = I_{R1} + I_{R2} + I_{R3}$, etc.

Resistance: (1) $R_T = \dfrac{1}{\dfrac{1}{R_1} + \dfrac{1}{R_2} + \dfrac{1}{R_3}}$ etc.

(2) $R_T = \dfrac{R}{R_n}$

(3) $R_T = \dfrac{R_1 \times R_2}{R_1 + R_2}$

Power: $P_T = P_{R1} + P_{R2} + P_{R3}$, etc.

PROBLEMS

Solve each of the following problems for the unknowns. Draw a schematic where none is given.

6-1. Two 60-Ω resistors are connected in parallel to a 6-volt source. Find I_T, I_{R1}, I_{R2}, R_T, and P_T.

6-2. Two resistors are parallel connected to a 40-volt source. One value is 100 Ω and the other is 200 Ω. Find I_T, R_T, I_{R1}, I_{R2}, P_{R1}, P_{R2}, and P_T.

6-3. Two resistors are parallel wired to a source. $I_T = 0.03$ A, $I_{R1} = 0.02$ A, and $R_1 = 10$ kΩ. Find E_T, I_{R2}, and R_2.

6-4. A 4.7-kΩ and a 5.6 kΩ resistor are parallel wired to a 12-V source. Find R_T and I_T.

6-5. A 3.3-Ω and a 3.9-Ω resistor are parallel wired to a 28-V source. Find I_T and R_T.

Use the schematic shown in Figure 6-9 for the following problems.

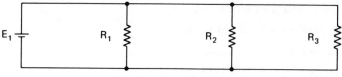

Figure 6-9

6-6. Each branch current is 0.030 A. The source is 20 V. Find I_T and R_T.

6-7. Branch currents are R_1 = 15 mA, R_2 = 30 mA, R_3 = 20 mA. The source voltage is 12 V. Find I_T and R_T.

6-8 I_{R1} = 80 mA, I_{R2} = 20 mA, I_{R3} = 0.04 A, E_T = 12 V. Find R_1, R_2, and R_3.

6-9. Determine the power for each resistor in Problem 8 and total circuit power.

6-10. Total circuit resistance is 450 mA. Source voltage is 24 V. Find R_T.

6-11. Source voltage is 15 V. R_1 = 27 Ω, R_2 = 33 Ω, and R_3 = 68 Ω. Find the current through each resistor, total circuit current, and power.

6-12. Four 120-V lamps are parallel wired in a circuit. The lamp ratings are 100, 150, 40, and 60 W. Find the current through each lamp and total circuit current.

QUESTIONS

6-1. A parallel circuit has a minimum of ____ loads.

    a. one

    b. two

    c. three

    d. four

6-2. Voltages across any load in a parallel circuit are

    a. equal to the source voltage

    b. less than the source voltage

    c. the sum of all voltage drops

    d. none of the above

6-3. Total current in any parallel circuit is

    a. less than source current

    b. greater than source current

    c. equal to the product of each load current

    d. equal to the sum of each load current

6-4. Increasing the number of loads in a parallel circuit

    a. increases total resistance

    b. decreases total resistance

    c. has no effect on total resistance

    d. drops the voltage at the source

6-5. Increasing the source voltage will

    a. increase current flow in the circuit

    b. decrease current flow in the circuit

    c. increase current flow through one branch of the circuit

    d. decrease resistance in the circuit

6-6. Decreasing load resistance in one load while maintaining the resistance value of the other loads will

    a. decrease current flow in the circuit

    b. increase current flow in all parts of the circuit

    c. change the applied voltage to the load

    d. increase current flow in that load only

6-7. Power in a parallel circuit is

    a. equal to each load's power consumption

    b. greater than each load's power consumption

    c. the sum of each load's power consumption

    d. not affected by each load's power consumption

6-8. Total circuit resistance in a parallel circuit is ____ the lowest single load resistance value.

    a. greater than

    b. less than

    c. equal to

    d. the sum of

6-9. The reciprocal of the number 100 is

    a. 100

    b. 10

    c. 0.1

    d. 0.01

6-10. Power in the load is _____ power in the source.

    a. equal to

    b. less than

    c. greater than

    d. has no bearing on

Combination Series-Parallel Circuits

Simple series or parallel circuits are not used exclusively in electrical devices. Very often the circuits used are composed of both series and parallel components. These circuits require analysis, also. Some basic procedures, if followed, will make the solution of series parallel circuits very simple.

IDENTIFICATION OF THE CIRCUIT

The first thing to do when analyzing any circuit is to determine the kind of circuit that it is. The basic electrical circuits are either series or parallel wired. The combination series-parallel circuit will also fall into one of these categories. The technician has to decide which of the two types most nearly describes the unknown circuit. Samples of every type of series parallel circuit would fill the pages of a textbook. Figure 7-1 offers some samples of these circuits. Examine each of these in order to determine the basic circuit configuration.

Circuit (a)

This circuit is basically a series circuit. Resistor R_1 is wired in series with the parallel combination of R_2 and R_3. This circuit is solved by first finding the equivalent resistance of R_2 and R_3. A two-step breakdown of the circuit is shown in Figure 7-2. The first step is to apply the laws relating to two parallel resistances. Change these into

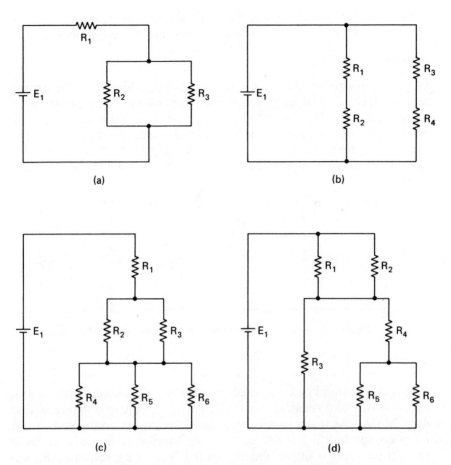

Figure 7-1. Typical series-parallel circuit configurations.

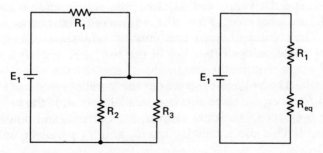

Figure 7-2. Resolution of a combination circuit into a basic series circuit by finding equivalent values for each branch.

a single equivalent resistance, R_e. Now the circuit values are used to identify specific voltages, currents, resistances, and power values.

Circuit (b)

The basic design of this circuit is a parallel combination. The circuit has two sets of resistors in series with each other. These are R_1 with R_2 and R_3 with R_4. Each set is connected in parallel to the power source. The first step in the solution of this circuit is to simplify the series-wired resistances. This is shown in Figure 7-3. Once the two equivalent resistances are found, the circuit is treated as a parallel circuit.

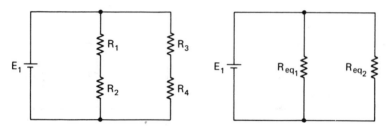

Figure 7-3.   Resolve each series branch into an equivalent value.

Circuit (c)

This is an excellent place to explain some circuit wiring. Figure 7-4(a) shows a tree-like resistive network. In this circuit it is assumed that each horizontal line has zero resistance. Anything connected at any place on the line is considered to be connected to the same point in the circuit. This is illustrated in part (b). This being the case, this circuit is basically a series circuit. It has two sets of parallel-wired resistors. These sets are R_2, with R_3 and R_4, R_5 with R_6. The circuit is best reduced to three resistances. This is shown in part (c).

When specific values are required, the subcircuits making up the network are analyzed. First, the equivalent resistances are determined. Then voltage drops and current values are calculated. The proof of these values often lies in the use of Kirchhoff's two laws. The last steps require the treatment of each of the three sections of this circuit. The voltage drop across the parallel combination of R_4, R_5, and R_6 is equal, since this is a parallel circuit. If the voltage drop and the resistance values are known, the currents and powers can be determined. This also is true for the R_2 and R_3 parallel combination.

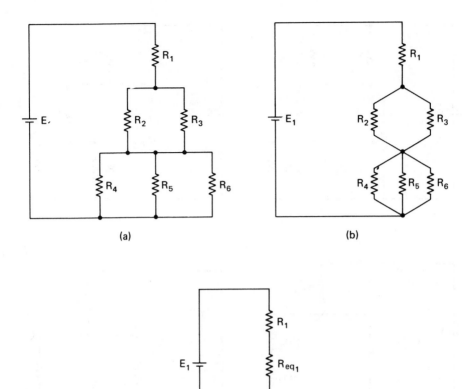

Figure 7-4. Resolution of a complex circuit into a simpler equivalent-value circuit.

Circuit (d)

This is the most complicated of the circuits used as examples. It has several sets of parallel and series combinations. The easiest way to break this circuit down is to decide whether it is a series or a parallel circuit. Once this decision is made, the treatment is the same as for the previous examples.

This circuit is more of a series circuit than it is a parallel circuit. The steps to reduce it are shown in Figure 7-5. Start by reducing R_1 and R_2 to an equivalent resistance labeled R_{e1}. Next reduce R_5 and R_6 to an equivalent resistance. Call this reduction R_{e2}. R_{e2} is series

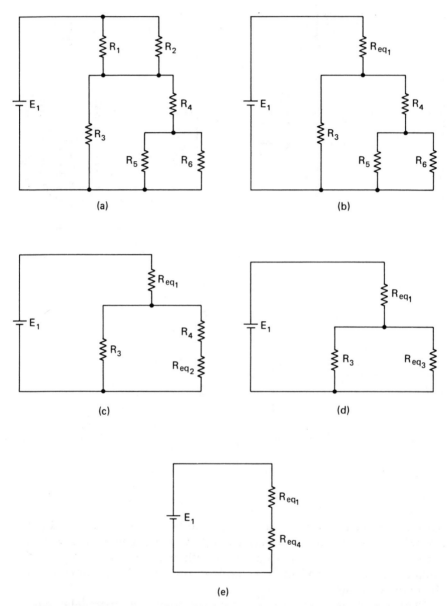

(a)

(b)

(c)

(d)

(e)

Figure 7-5. Each section of the circuit is reduced to an equivalent value as the circuit is resolved.

wired to R_4. These, too, are reduced to one resistance identified as R_{e3}. The next step is to reduce the parallel circuit R_3 and R_{e3} to a single value of resistance. This is identified as R_{e4}. This procedure results in a two-resistance series circuit. Values can be identified for the two resistor values. Once this is determined, the circuit with its specific values can be reconstructed. Values for each component can easily be found using Ohm's, Watt's and Kirchhoff's laws.

SOLUTION OF CIRCUIT VALUES

This section of the chapter uses samples of complex electrical circuits. Examples of how to find values of voltage, current, resistance, and power for each component are given. The procedures used are very similar to the procedures given in the previous section. Always attempt to identify whether the circuit is series or parallel wired as the starting point. The next step is to simplify the circuit. Then the rules for circuit analysis are applied.

Figure 7-6 illustrates the steps involved in complex circuit analysis. The first step is to determine whether this circuit is basically series or parallel wired. The clue here is the connections to R_1. It is wired in series between the power source and the rest of the circuit. Use the illustration to visualize how the simplification of the circuit takes place.

Start at a point farthest away from the source. Resistors R_5 and R_6 are series connected. These two resistors are parallel wired to resistor R_4. Apply the rule for resistors in series.

$$R_{eq} = R_5 + R_6 = 200 + 200 = 400 \ \Omega$$

The equivalent resistance of 400 Ω is in parallel with another 400 Ω resistance, R_4. Apply the rule for parallel resistances.

$$R_{eq} = \frac{R}{R_n} = \frac{400}{2} = 200 \ \Omega$$

Therefore, the equivalent resistance of R_4, R_5, and R_6 is 200 Ω. This 200-Ω equivalent resistance is in series with R_3. Apply the rule for for resistors in series.

$$R_{eq} = R_3 + R_{4,5,6 \, eq} = 200 + 200 = 400 \ \Omega$$

The equivalent resistances of R_3, R_4, R_5, and R_6 are connected in parallel to R_2. The resistance of this combination is found by applying the parallel resistance rule.

$$R_{eq} = \frac{R}{R_n} = \frac{200}{2} = 200 \ \Omega$$

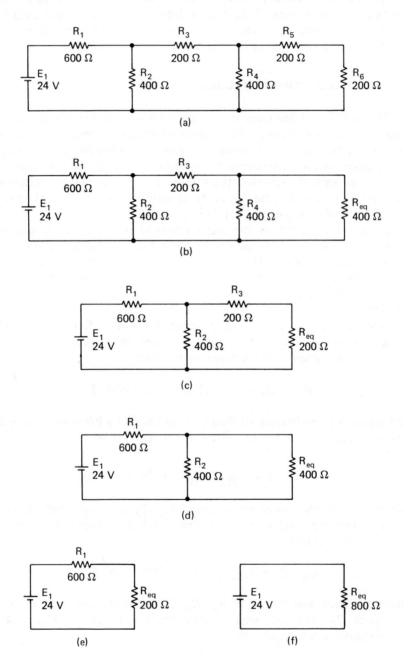

Figure 7-6. Solution of a complex series-parallel circuit.

The equivalent resistance of all resistors except R_1 is series connected to R_1. Find the total resistance by adding each resistor's ohmic value. The answer is the total circuit resistance.

$$R_T = R_1 + R_{eq} = 600 + 200 = 800 \; \Omega$$

The total resistance in the circuit is found to be 800 Ω. Once this is determined and the source voltage is known, the total circuit current is determined. Ohm's law is used to find circuit current.

$$I_T = \frac{E_T}{R_T} = \frac{24}{800} - 0.03 \; A$$

Now refer to Figure 7-7. This is the original circuit shown in the example. Ohm's and Watt's law's triangles are added beside each

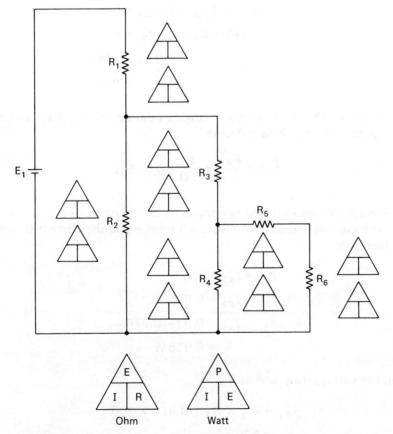

Figure 7-7. Use of Ohm's law and Watt's law triangles to solve for circuit values.

component in the circuit. All information given on the original schematic is included in each triangle. This information is essentially the

The only component whose current flow is known is R_1. This is
because R_1 is series wired between one side of the source and the
rest of the resistors. There is only one path for current flow through
this part of the circuit. Therefore, 100% of circuit current, or 0.03 A,
flows through this resistor. The resistance and current values for this
resistor are known. Now the voltage drop across R_1 is found.

$$E_{R1} = I_{R1} \times R_1 = 0.03 \times 600 = 18 \text{ V}$$

Resistors R_1 and R_2 are series wired to the 24-V source. Apply
Kirchhoff's law for voltage in a closed loop. Ignore the balance of the
circuit for the moment.

$$E_{R1} + E_{R2} + E_{\text{source}} = 0$$
$$+18 + E_{R2} + (-24) = 0$$
$$E_{R2} = -18 + 24$$
$$= 6 \text{ V}$$

Add the 6-V value to the triangles relating to R_2. The current
through this resistor is found next.

$$I_{R2} = \frac{E_{R2}}{R_2} = \frac{6}{400} = 0.015 \text{ A}$$

Kirchhoff's current law is applied to the junction of R_1, R_2, and
R_3. The current entering the junction must equal the current leaving
the junction.

$$I_{R2} + I_{R3} = I_{R1}$$
$$0.015 + I_{R3} = 0.03$$
$$I_{R3} = 0.03 - 0.015$$
$$= 0.015 \text{ A}$$

Find the voltage drop across R_3.

$$I_{R3} = I_{R3} \times R_3 = 0.015 \times 200$$
$$= 3 \text{ V}$$

Find the voltage drop across R_4. The combination of R_3 and R_4 is wired in parallel across R_2. The voltage drop across R_2 is equal to the drop across the series-wired R_3 and R_4. The voltage drop across R_4 is the difference between the applied voltage and the drop across R_3.

$$E_{R4} = E_{R2} - E_{R3}$$
$$= 6 - 3$$
$$= 3 \text{ V}$$

This uses a variation of Kirchhoff's law. The variation states that the voltage drops across each component in a series-wired circuit add to equal the source voltage. In this circuit the voltage drop across R_2 is considered to be the source for the analysis.

Next find the current through R_4. This current is

$$I_{R4} = \frac{E_{R4}}{R_4} = \frac{3}{400} = 0.0075 \text{ A}$$

The current through R_5 is now found by use of Kirchhoff's current law.

$$I_{R5} + I_{R4} = I_{R3}$$
$$I_{R5} = I_{R3} - I_{R4}$$
$$= 0.015 - 0.0075$$
$$= 0.0075 \text{ A}$$

The voltage drop across R_5 is now determined.

$$E_{R5} = I_{R5} \times R_5$$
$$= 0.0075 \times 200$$
$$= 1.5 \text{ V}$$

The voltage drop across R_6 is the difference between the total voltage across the series combination R_5, R_6 and the voltage drop across R_5.

$$E_{R6} = (E_{R5} + E_{R6}) - E_{R5}$$
$$= 3 - 1.5$$
$$= 1.5 \text{ V}$$

The current through R_6 is found by use of Ohm's law. It may also be found by deductive reasoning. If R_5 and R_6 are series wired, the current through R_5 is equal to the current through R_6.

This completes the determination of voltage, current, and resistance for each component in this circuit. All that is left is to determine the power consumption for each of the components. Since voltage and current values are known, it is fairly simple to do this. Apply one section of the power formula.

$$P_T = E_T \times I_T$$
$$= 24 \times 0.03$$
$$= 0.72 \text{ W}$$
$$P_{R1} = E_{R1} \times I_{R1}$$
$$= 18 \times 0.03$$
$$= 0.54 \text{ W}$$

Continue this procedure for each component in the circuit. When completed, total power should equal the power consumed by each component in the circuit.

A review of some alternative methods of finding circuit values is in order before proceeding to another problem. These analytical methods will save time as well as help direct the thinking process. Use the following statements to relate Kirchhoff's and Ohm's laws.

1. If R_5 and R_6 are in series, then
    a. The current through R_5 equals the current through R_6.
    b. The voltage drop across R_5 is equal to the voltage drop across R_6.
    c. Power consumed by R_5 is equal to power consumed by R_6.
2. If the series combination $R_5 + R_6$ is equal to the value of R_4, then
    a. Current through R_4 is equal to the current through R_5 and R_6.
    b. The voltage drop across R_5 and R_6 is equal to the voltage drop across R_4.
    c. The power consumed by R_4 is equal to the power consumed by $R_5 + R_6$.
3. If the ohmic value of R_2 is equal to the total resistance of the circuit using R_3, R_4, R_5, and R_6, then

   a. Current flow through R_2 is equal to current flow through the resistive network.

   b. The voltage drop across R_2 is equal to the voltage drop across the equivalent network.

   c. Power consumed by R_2 is equal to the total power consumption of the resistive network R_3 through R_6.

These concepts will aid in the identification of circuit values.

A second series parallel circuit is shown in Figure 7-8. The pro-

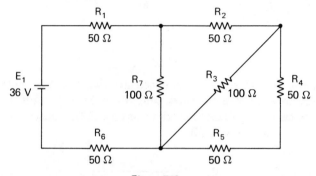

Figure 7-8

cedure for analysis of this curcuit is similar to that shown for the previous circuit. The initial step is to redraw the circuit. This aids in the visualization of the circuit. It also is the first step in determining whether the circuit is basically a series or a parallel circuit. The redrawn circuit is shown in Figure 7-9. This circuit is basically a series circuit. It has two parallel-connected branches. One of these is the

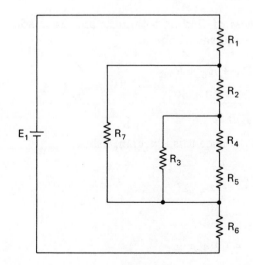

Figure 7-9. Another method of illustrating the circuit shown in Figure 7-8.

combination of R_4, R_5, and R_3. The second is the combination of R_3, R_4, R_5, R_2, and R_7. Reduce the simplest parallel combination first. R_4 and R_5 are series connected. Their total resistance is 100 Ω. This combination is in parallel with R_3. The equivalent resistance of this resistive network is 50 Ω.

$$
\begin{aligned}
R_{eq} &= \frac{(R_4 + R_5) \times R_3}{R_4 + R_5 + R_3} \\
&= \frac{100 \times 100}{200} \\
&= \frac{10,000}{200} \\
&= 50 \ \Omega
\end{aligned}
$$

Resistor R_2 is series connected to this resistive network. The total resistance of the network and R_2 is 100 Ω. This 100-Ω equivalent resistance is parallel wired across R_7. The equivalent resistance of this second network is 50 Ω.

$$
\begin{aligned}
R_{eq} &= \frac{(R_2 + R_{eq1}) \times R_7}{R_2 + R_{eq1} + R_7} \\
&= \frac{(50 + 50) \times 100}{50 + 50 + 100} \\
&= \frac{10,000}{200} \\
&= 50 \ \Omega
\end{aligned}
$$

The next step in this process of finding total resistance is to add the values of R_1 and R_6 to the equivalent circuit.

$$
\begin{aligned}
R_T &= R_1 + R_6 + R_{eq2} \\
&= 50 + 50 + 50 \\
&= 150 \ \Omega
\end{aligned}
$$

Solve for total circuit current once this is accomplished.

$$
\begin{aligned}
I_T &= \frac{E_T}{R_2} \\
&= \frac{36}{150} \\
&= 0.24 \ \text{A}
\end{aligned}
$$

The balance of the values for each circuit component are found at this time. There is only one current path through both R_1 and R_6. All circuit current flows through these resistors. Draw Ohm's law triangles next to each of the resistors in the circuit. Add the current values for R_1 and R_6. Once this is done the voltage drop across each of these resistors may be determined.

$$E_{R1} = I_{R1} \times R_1$$
$$= 0.24 \times 50$$
$$= 12 \text{ V}$$
$$E_{R6} = I_{R6} \times R_6$$
$$= 0.24 \times 50$$
$$= 12 \text{ V}$$

Another way of doing this is to use the rule of substituting equal values.

$$R_1 = R_6$$
$$E_{R1} = E_{R6}$$
$$12 = 12$$

In other words, if resistor values in a series circuit are equal, the voltage drops across each resistor also will be equal.

The voltage drop across the balance of the circuit is found by use of Kirchhoff's voltage law. The equivalent circuit R_{eq2} is series connected to the source, R_1 and R_6. Use Kirchhoff's law:

$$E_{source} + E_{R1} + E_{Req} + E_{R6} = 0$$
$$36 - 12 + E_{Req} - 12 = 0$$
$$36 - 24 = E_{Req}$$
$$12 = E_{Req}$$

Voltage and resistance values for R_7 are now known. Find current through R_7.

$$I_{R7} = \frac{E_{R7}}{R_7} = \frac{12}{100} = 0.12 \text{ A}$$

This information is used to determine the current through R_2. Kirchhoff's current law is applied at this point to the junction of R_1, R_2, and R_7. The currents flowing through R_2 and R_7 are equal to the current flowing through R_1.

$$I_{R1} = I_{R2} + I_{R7}$$
$$0.24 = I_{R2} + 0.12$$
$$0.24 - 0.12 = I_{R2}$$
$$0.12 = I_{R2}$$

The voltage drop across R_2 can now be found.

$$E_{R2} = I_{R2} \times R_2$$
$$= 0.12 \times 50$$
$$= 6 \text{ V}$$

If there is a 6-volt drop across R_2, the balance of the voltage drop across R_7 must develop across the parallel network R_3 and $R_4 + R_5$.

$$E_{R3} = E_{R7} - E_{R2}$$
$$E_{R3} = 12 - 6$$
$$= 6 \text{ V}$$

This provides the information necessary to find the current through R_3.

$$I_{R3} = \frac{E_{R3}}{R_3}$$
$$= \frac{6}{100}$$
$$= 0.06 \text{ A}$$

The resistive path of R_3 is equal to the resistive path of R_4 and R_5. The currents through each path are equal to one half of the current through R_2. This is proven by use of Kirchhoff's law.

$$I_{R3} + I_{R4} = I_{R2}$$

$$0.06 + 0.06 = 0.12$$

The voltage drops across R_4 and R_5 are found by use of Ohm's law.

$$E_{R4} = I_{R4} \times R_4$$
$$= 0.06 \times 50$$
$$= 3 \text{ V}$$

$$E_{R5} = I_{R5} \times R_5$$
$$= 0.06 \times 50$$
$$= 3 \text{ V}$$

A little deductive reasoning will provide these same voltage drop values without using Ohm's law. If any resistors are connected in series and their ohmic values are equal, then the voltage in the circuit drops equally across each of the resistors. In this case, the 6 volts applied to this section of the circuit is divided equally across the two resistances. The end result is a 3-volt drop across resistors R_4 and R_5.

SUMMARY

Series-parallel circuits have to be simplified in order to determine values of resistance, current, voltage drop, and power for each component. To solve for the unknown quantities, the circuit first must be identified as either a basic series or a basic parallel circuit. Redrawing the circuit using either vertical or horizontal lines will often make the circuit easier to follow.

Simplification of the circuit to find total resistance usually starts with the smallest part of the network. This part becomes an equivalent circuit. The equivalent value is used to determine the next smallest set of values. This procedure is continued until one value of resistance is known. This value is the equivalent of all resistances in the circuit.

Once the total voltage and resistance are known, Ohm's law is applied to find total current.

Circuit analysis continues as the values of voltage, resistance, and current are identified for each component. Use of both of Kirchhoff's laws as well as Ohm's law is required when solving circuit problems.

PROBLEMS

Refer to Figure 7-10 for the following problems.

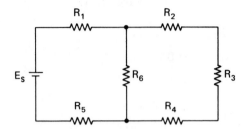

Figure 7–10

7-1. E_s = 45 V, R_1 = R_5 = 150 Ω, R_2 = R_3 = R_4 = 100 Ω, R_6 = 300 Ω; find R_T, I_T, I_{R3}, I_{R6}, E_{R6}.

7-2. E_s = 100 V, R_1 = 1.5 kΩ, R_2 = 1.8 kΩ, R_3 = 200 Ω, R_4 = 2.2 kΩ, R_5 = 2.4 kΩ, R_6 = 3 kΩ; find R_T, I_T, I_{R3}, I_{R6}, E_{R3}.

7-3. E_s = 900 V, R_1 = 30 kΩ, R_2 = R_3 = R_4 = 20 kΩ, R_6 = 60 kΩ; find R_T, I_T, I_{R3}, I_{R6}, E_{R6}, P_{R1}, P_T.

Refer to Figure 7-11 for the following problems.

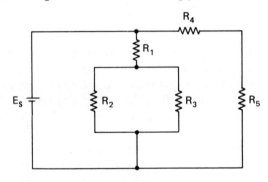

Figure 7–11

7-4. R_1 = 150 Ω, R_2 = 200 Ω, R_3 = 600 Ω, R_4 = 100 Ω, R_5 = 200 Ω, I_T = 3 A; find E_T, E_{R1}, E_{R2}, E_{R3}, E_{R4}, E_{R5}, I_{R5}, P_{R4}, P_{R2}, P_T.

7-5. E_s = 600 V, E_{R4} = 250 V, E_{R3} = 200 V, I_{R1} = 1 A, R_2 = 250 Ω, I_T = 1.5 A; find R_T, R_1, E_{R2}, I_{R2}, I_{R3}, R_3, I_{R4}, R_4, I_{R5}, R_5, P_T, P_{R4}.

Refer to Figure 7-12 for the following problems.

7-6. R_1 = 60 Ω, R_2 = R_4, I_{R1} = 10 A, E_{R4} = 4 kV; find E_{R2}, E_{R3}, E_{R1}, I_{R3}, E_T, I_T, R_T.

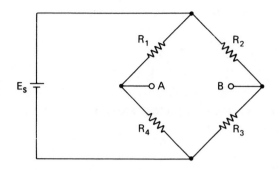

Figure 7-12

7-7. In problem 7-6, is there a current flow between points A and B? If there is, explain why. If there is no current flow, explain why.

7-8. R_1 = 330 Ω, R_2 = 470 Ω, R_3 = 56 Ω, E_s = 100 V. What value of resistance is necessary at R_4 in order to have the voltage at point A equal to the voltage at point B?

QUESTIONS

7-1. In Figure 7-1(a) voltage across R_2 is ____ voltage across R_3.

   a. less than

   b. equal to

   c. more than

   d. independent of

7-2. In Figure 7-1(a), R_2 = 100 Ω and R_3 = 50 Ω. Current through R_2 is ——— than R_3.

   a. less than

   b. equal to

   c. more than

   d. independent of

7-3. In Figure 7-1(a), voltage drop across R_1 is ____ R_2 and R_3.

   a. less than

   b. equal to

   c. more than

   d. independent of

7-4. In Figure 7-1(b), current through R_1 is _____ current through R_2.

   a. less than

   b. equal to

   c. more than

   d. independent of

7-5. In Figure 7-1(b), voltage across R_1 is _____ voltage across R_3.

   a. less than

   b. equal to

   c. more than

   d. independent of

7-6. In Figure 7-1(b), the voltage drop across $R_1 + R_2$ is _____ the voltage drop across $R_3 + R_4$.

   a. less than

   b. equal to

   c. more than

   d. independent of

7-7. In Figure 7-1(d), the voltage drop across R_1 is _____ the voltage drop across R_2.

   a. less than

   b. equal to

   c. more than

   d. independent of

7-8. In Figure 7-1(d), current through R_1 is _____ the current through R_3.

   a. less than

   b. equal to

   c. more than

   d. independent of

7-9. In Figure 7-1(c), the voltage drop across R_4 and R_5 is _____ the voltage drop across R_6.

   a. less than

   b. equal to

   c. more than

   d. independent of

7-10. In Figure 7-12, if all resistor values are equal, the voltage drop at *A* is ___ the voltage drop at *B*.

    a. less than

    b. equal to

    c. more than

    d. independent of

8

Magnetism

Some of the most fundamental scientific laws of our world are those relating to magnetism. Magnetism in both natural form and electrical form has shaped our world. Many major technological advances are based on magnetic principles. Magnetism helps produce the large quantities of electrical power required by home, business, and industry. The study and understanding of magnetic forces and their applications is very important if one is to really know how electrical devices work.

HISTORY

Magnetic properties in natural form were discovered early in the history of humanity. Certain stones were found to align themselves along a north-south axis when the stones were suspended on a string. These stones contained minerals that had natural magnetic properties. They were called *lodestones* or *magnetite*. The first recorded use of lodestone was about 800 years ago. These stones were used as aids in the navigation of ships. Today we use a very similar device called a compass.

MAGNETS AND LINES OF FORCE

The needle of the compass points to the earth's magnetic north pole. This is because the earth itself acts like a huge magnet. Every magnet has two areas called *poles*. One is identified as the north pole, and the other is the south pole. Studies of magnetic properties have found that the polar areas are those where the magnetic force leaves the magnet and returns to it. Each magnetic pole has its corresponding pole. In other words, for each north pole there is a matching south pole on the magnet.

Magnetic poles are present owing to magnetic lines of force. Under normal conditions these lines of force are neither seen nor felt. They can, however, be illustrated using a magnet, a piece of paper, and iron filings. The paper is placed on top of the magnet. The iron filings are then sprinkled on the paper. The filings will form a pattern similar to that shown in Figure 8-1. Several things are apparent when one studies this action:

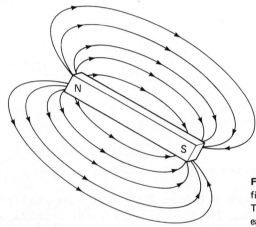

Figure 8-1. A magnetic force field is formed around all magnets. The field strength is greatest at each pole.

1. The lines of force are concentrated at the ends of the magnet.
2. The lines of force do not touch each other.
3. The lines of force go from one pole to the other pole.

Further study of the lines of force around a magnet show that if the magnet is turned on its side the lines of force appear to act in the same manner. One conclusion from this is that the force field completely surrounds the magnet. It does not work in only one direction. This is illustrated in Figure 8-2, which shows one end of a bar of magnetic material. The lines of force are all around the end. These lines of force leave one pole of the magnet and go to the opposite pole of the same magnet.

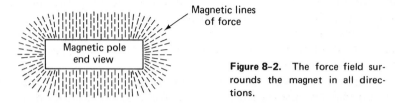

Magnetic lines of force

Figure 8-2. The force field surrounds the magnet in all directions.

The study of magnetic fields resulted in laws relating to the interaction of two magnetic fields. Each individual magnet has its own force field. If two magnets are placed near each other, one magnet's force field acts on the other magnet's force field. This is illustrated in Figure 8-3. When one magnet's north pole is placed near the other

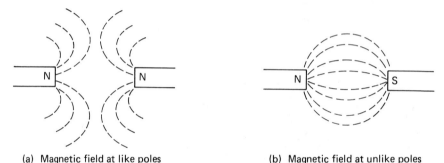

(a) Magnetic field at like poles   (b) Magnetic field at unlike poles

Figure 8-3. Magnetic fields from one magnet act upon any other magnet.

magnet's south pole, as shown in Figures 8-4(a) and 8-4(c), the two fields complement each other. The lines of force from one magnetic pole are attracted to the other magnet's magnetic pole. Note also that some of the lines of force still go between each magnet's poles. Now the two magnet's fields complement each other. The lines of magnetic force reinforce the total magnetic field. This situation occurs when opposite magnet poles are placed near each other.

The other two parts of Figure 8-4 illustrate the effect of placing two similar magnetic poles near each other. When this occurs, the magnetic lines of force do not cross from one magnet to the other magnet. These fields tend to oppose and move away from each other. This is true when either the two north poles or the two south poles are placed near each other.

From these illustrations come some rules about magnetism:

4. Like magnetic poles repel each other's lines of force.

5. Opposite magnetic poles attract each other's lines of force.

6. Magnetic lines of force do not cross, but tend to be close to other lines of force.

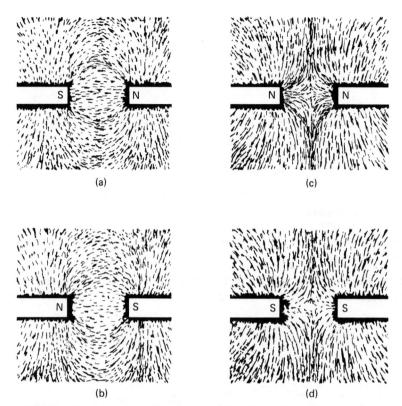

Figure 8-4. The interaction of various magnetic fields. Each pole will react with any other pole's magnetic field as shown.

MAGNETIC MATERIALS

There are two kinds of easily magnetized materials. They are classified as either magnetically *hard* or magnetically *soft*. The difference is in the ability of the material to retain its magnetic properties after the force that produces the magnetism is removed. Pure iron, for example, is a magnetically soft material. If a magnet is placed near a piece of pure iron, the iron will take on the magnetic properties of the magnet. Thus, it will form its own magnet. This also tends to change the shape of the magnetic force field. This is illustrated in Figure 8-5. The insertion of the smaller soft iron bar makes it a magnet. It also bends the lines of force from the larger magnet, thus changing the shape of the force field. Removal of the smaller soft iron bar returns the larger magnet's force field to its original shape. The soft iron bar will lose its magnetic properties at the same time.

If an iron washer is placed in the magnetic field, the lines of force

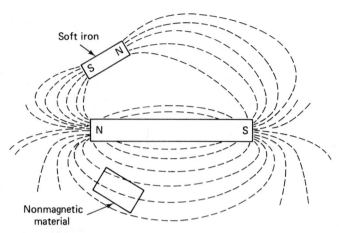

Figure 8-5. The shape of the magnetic force field may be directed by use of additional magnetic fields.

will be directed *around* the hole in the middle of the washer. This is illustrated in Figure 8-6. The lines of force tend to take the easier

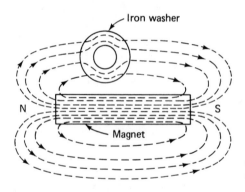

Figure 8-6. A metal iron washer will direct the forces around a specific area in the field.

path. In this case, as in most, it is easier for the lines of force to travel through the iron washer than through the air. Therefore, the center of the washer will have few, if any, lines of force present.

Other materials, mainly iron alloys, are magnetically hard. These materials will become permanently magnetized when they are subjected to a magnetic force field. In addition to iron alloys, some ceramic materials are also used as permanent magnets. One common commercial magnetic material is *alnico*, an alloy of aluminum, nickel, cobalt, and iron. Alnico magnets are capable of producing very strong magnetic fields in relation to the physical size of the magnet. Plastic materials are mixed with magnetic materials and formed into a variety of shapes for use as magnets. One application of this in the home is found in the strip magnets used to hold refrigerator doors closed.

SHAPES OF MAGNETS

Permanent magnets are manufactured in a variety of shapes. Each shape produces its own magnetic field. Using the principle that opposite magnetic fields attract each other, the closer together the poles are, the stronger the force field outside the magnet will be. This is because the lines of magnetic force are more concentrated.

Two basic shapes of permanent magnets are illustrated in Figure 8-7. We have shown the bar magnet in an earlier illustration. Shown

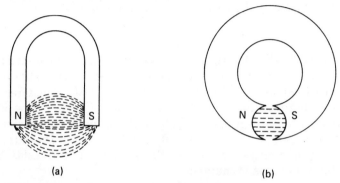

(a)   (b)

Figure 8-7. Common shapes for magnets are the horseshoe and the ring magnet. Other shapes are also used, but not illustrated.

here are other shapes. Figure 8-7(a) is a horseshoe magnet. The concentration of the magnetic field outside of the magnet is between the pole pieces. Part (6) shows a ring magnet. This type of magnet has uniform lines of force throughout its circular frame. It lends itself well to use in motors and electrical meters.

Another shape (not illustrated) is the round magnet found in a loudspeaker. Actually, the magnets used today are found in a variety of physical shapes. The final form depends upon the application. All of these magnetic forms have one thing in common, the pair of poles. Every magnet will have one north and one south pole, regardless of the shape of the magnet.

MAGNETIC TERMS AND UNITS

People have always seemed to need a means of measuring quantities. Communication between people requires, among other things, use of a common language. The discussion of magnetism is easier if one also uses a common language. There are a large number of terms used when describing magnetic properties. Some of the more common terms are presented and explained in this section. All terms use the SI (metric) system of measurement.

Magnetic Flux

Magnetic flux is the term used to describe the entire magnetic force field surrounding any magnet. The *weber* (Wb) is the unit used to represent the flux of the magnet. A number is also used to help describe the flux field. Higher numbers indicate a larger flux field density. In other words, as the descriptive numbers increase, there are more lines of force, or flux, in the field. This term is very large. It is seldom used when calculating flux. A more common method uses a fraction of the weber. The microweber (μWb) is most often used. One microweber represents 100 lines of flux in the magnetic field.

Flux Density

Flux density is the description of the number of magnetic lines of force per unit area. The SI system uses one weber per square meter as the point of reference. This unit is called the *tesla*. The Greek letter Φ (phi) is used to represent the flux density of a magnetic field.

Flux density of a magnetic field is the number of magnetic lines of force in a given area. The formula $\beta = \Phi/A$ is used for determining flux density in a magnetic field. In this formula, β is in teslas, Φ (phi) is in webers, and A is in meters.

Permeability

Permeability relates to the ability of a material to be magnetized. If any substance is capable of becoming highly magnetized, it is said to have a high permeability. Permeability is measured by the number of magnetic lines of force produced in the substance by a fixed magnetic force. This is found by use of the formula $\mu = B/H$, where μ is the permeability value, B is in webers, and H is the number of ampere-turns per meter.

Reluctance

The opposition to magnetic lines of force is called reluctance. Materials that are easily magnetized have a low reluctance. Materials that are difficult to magnetize have a high reluctance. Reluctance is expressed in ampere-turns per weber. The formula $\mathcal{R} = \mathcal{F}/\Phi$ is used to calculate reluctance. Reluctance is the reciprocal of permeance, and reluctivity is the reciprocal of permeability.

Retentivity

Retentivity is the measure of how well a material is able to hold its magnetic properties after it is exposed to a magnetizing force. This is measured after the magnetizing force is removed. The terms *hard* and *soft* are used to describe this. A magnetically hard material has excellent retention capabilities. The magnetically soft material loses its magnetic properties quickly after the magnetizing force is removed. Another way of stating this is that permanent magnets have a high retentivity value and temporary magnets have a low retentivity value.

Residual Magnetism

This term relates to the amount of magnetic field strength left in any magnetic material after the magnetizing force is removed. Magnetically hard materials have a high residual magnetism. Residual magnetism is low in magnetically soft materials. The amount of residual magnetism is measured in terms of the flux density at the poles of the magnet.

Permeance

Permeance is the reciprocal of reluctance. This refers to how easily the magnetic field passes through the magnetic material. In the SI system, permeance is figured by dividing the number of webers by the ampere-turns of the magnet. The formula for permeance is

$$\text{permeance} = \frac{1}{\Re}$$

Ampere-turns

The term ampere-turns is used to describe the magnetizing force required to produce a magnet by means of electricity. This is illustrated in Figure 8-8. The reference for this term is a coil that is wound on a nonmagnetic form. The form could be cardboard or even air. The reason for this is to eliminate any effect of the core on the resulting magnet. An electrical current is passed through the coil of wire. The amount of electrical current is measured as this occurs. The number of complete turns of wire that form the coil is multiplied by the

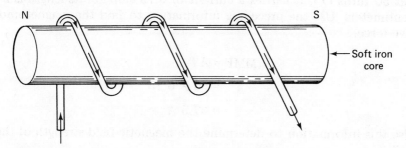

Figure 8-8. A magnetizing force's strength is described by use of the term ampere-turns. A coil wound on a nonmagnetic form is used as a reference.

amount of current flowing. This produces a number that is considered to be the magnetizing force of the coil. This force is called the magnetomotive force (MMF). It is measured in ampere-turns. The symbol used to describe the number of ampere-turns is MMF. Two electromagnetic circuits are illustrated in Figure 8-9. One shows a coil that

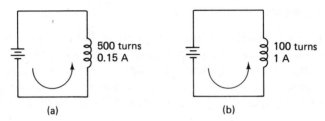

Figure 8-9. Ampere turns are determined by the number of turns of wire on the form and the current flowing through these turns of wire.

has 100 turns of wire. The measured electrical current is 1 ampere. The magnetomotive force for this coil is 100 × 1, or 100 MMF. Part (b) shows a coil that has 500 turns of wire. The measured electrical current flowing through the coil is 0.15 amperes. The MMF of this coil is 500 × 0.15, or 22.5 MMF. In summary, MMF is equal to 1 turn of wire on the coil and 1 ampere of current flowing through the coil.

MAGNETIC FIELD STRENGTH

Magnetic field strength is measured in units of ampere-turns per meter. The letter H is used for this measurement. The formula $H = \mathcal{F}/l$ is used to determine field strength. H is the magnetic field strength, $\mathcal{F}$ is the magnetomotive force in ampere-turns, and l represents the length in meters of the coil form. The answer is given in amperes per meter.

An example of the use of this formula is as follows: A coil of wire has 50 turns (T). It carries a current of 0.75 ampere. Its length is 25 centimeters. Use the preceding information to find the magnetomotive force.

$$MMF = AT$$
$$= 50 \times 0.75$$
$$= 37.5 \text{ A}$$

Use this information to determine the magnetic field strength of the coil.

$$H = \frac{\text{MMF}}{l}$$

$$= \frac{37.5}{25 \times 10^{-2}\,\text{m}}$$

$$= 150 \text{ A/m}$$

It is interesting to be able to draw some conclusions from the use of these formulas. Consider the effect of the number of turns of wire on the amount of magnetic field. As the number of turns of wire increases, the magnetic field also increases. The same is true about the amount of current flowing. If this increases owing to a larger supply source or to the size of the wires used to wind the coil, the size of the magnetic field also increases. The overall length of any coil also affects its magnetic strength. As the length is increased, the coil has a lower magnetic force per unit of measure. All these factors are used in determining the quality of any electromagnetic device.

MAGNETIC HYSTERESIS

Most magnetic materials have a tendency to keep some magnetic effect after the forces used to magnetize them have been removed. A magnetically soft material is used in the production of electromagnetic devices. Typical materials include soft iron, hard steel, and ferrite. Ferrite is a powdered iron material that is formed into the shape of a magnetic core. Each of these three materials exhibits different magnetic qualities. These are illustrated in Figure 8-10. The vertical

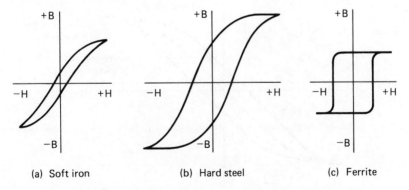

(a) Soft iron   (b) Hard steel   (c) Ferrite

Figure 8-10. Magnetization curves for typical magnetic materials.

lines on the chart represent the flux density of the magnet. The symbols + and − are used to indicate the ability of applying a magnetizing force in either direction. This means that the polarity of the force may be reversed. The horizontal line represents the field strength of the magnet created in this manner. Here, too, the symbols + and − are used to show a polar reversal of the magnet due to the effects of the applied force.

These three illustrations show that the forces of magnetism are not instantaneous. They require a little time to magnetize each of the materials. The left side of each curve represents the amount of magnetism produced in the core when the magnetizing force is applied to the core. The curve on the right side of each figure shows the time delay that occurs when the magnetizing force is removed. Note that each curve is symmetrical. Also consider the shaded area in each of the curves. Soft iron has the smallest shaded area. This indicates that soft iron cores maintain magnetic properties for the shortest period of time.

Also note the curve of the ferrite core. This core has rather clearly defined vertical lines. This means that the magnetic effect is turned "on" quickly. It is also turned "off" quickly. This capability is useful in computers and other data-processing devices where a hangover of the magnetic field will affect the subsequent operation of the device. Computers require a very positive magnetic field with little or no time lag. Ferrite core materials in electric magnets provide this type of reaction. Iron cores have a relatively long carryover time and cannot be used successfully for this purpose.

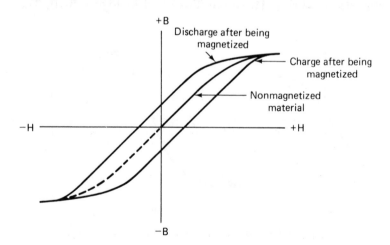

Figure 8-11.   Magnetic hysteresis causes a delay in the time required to either magnetize or demagnetize a material.

This hysteresis curve information relates to the frequency of any electrical waves being processed by the coils. A gradual hysteresis curve, such as illustrated in Figure 8-11, responds slowly to changes in electrical waves. These types of cores are best used at lower frequencies or changes. These are best illustrated by identifying power line frequenceis as low frequencies. Faster changes occur as the rate of change, or frequency, increases. Radio, television and data processing require fast changes at much higher frequencies. These systems would use ferrite core materials for any coils associated with these higher frequencies.

Each magnetic material has a set of hysteresis curves. Persons working in the design of electromagnets will have a need for this information. Others may wish to know what they mean, but will have very little use for the information.

DEGAUSSING

Magnetically hard materials are very difficult to demagnetize. One method that has proved successful is to apply a rapidly changing magnetic field to the magnet. This is done with a device called a degaussing coil. This coil consists of many turns of wire. It is powered by an alternating polarity electrical current such as used in the home. The coil is placed near the magnetic device and then the coil is slowly taken away. The effect on the magnetically hard material of this changing electromagnetic field neutralizes the magnet. This minimizes any residual magnetism in the material.

Common use of this principle is found in any color television set. The color picture tube is affected by magnetic fields. The steel frame supporting the picture tube acts like a magnet. It is magnetically hard. Most color television sets have a built-in degaussing coil that is used to neutralize this magnetic effect. The circuit only works during the first minute or so that the set is turned on. Television technicians use a portable degaussing coil to neutralize the color picture tube's magnetic field on a new installation.

Another application of the degaussing coil is the removal of residual magnetism on a tape recording head. This is required on industrial, broadcast, and home quality tape recorders. The magnetically hard tape heads build up a magnetic field. This field affects the quality of the recording. It must be neutralized periodically to provide true fidelity of the information recorded on the tape.

The term *gauss* is used in other systems of measurement to indicate the flux density of a magnet. This term is replaced in the SI system by the tesla. The term *degaussing* is a carryover from the earlier use of this term.

SUMMARY

There are several important concepts and rules used when study-
ing magnetism. These may be summarized in the following manner:
Each magnet has two poles. These are identified as the north pole
and the south pole.

Like magnetic poles repel each other, because the lines of magnetic force with similar polarities repel each other.

Opposite magnetic poles attract each other.

Magnetic materials include cobalt, iron, nickel, and steel.

Magnetic materials are classified as either magnetically hard or
soft. A hard material retains its magnetic properties after any magnetizing force is removed and makes a good permanent magnet. Soft materials make temporary magnets because they do not retain their magnetic properties.

The weber is the unit of magnetic flux. It indicates the number
of lines of flux in the magnetic field.

The tesla is the unit of magnetic flux density. The letter β is used
to represent flux density for a specific area of one square meter.

Permeance is the opposite of reluctance. It indicates how easily a
magnetic force passes through a material. It is determined by dividing
the number of webers by the ampere-turns of the coil.

Reluctance is opposition to a magnetizing force. Low reluctance
materials are easily magnetized.

Retentivity is the measure of how well a substance retains magnetism once the magnetizing force is removed.

Permeability is the ability of a given material to be magnetized. It
is represented by a number. The higher the number is, the greater the
ability of the substance to be magnetized. The basis for reference is
related to air. Air has a permeability factor of 1. Materials with high
permeability numbers are easily magnetized.

Ampere-turns refer to the number of turns of wire used to form a
coil and the amount of electrical current flowing through the coil.

Magnetic field strength is the measure of the number of ampere-
turns per meter. The letter H represents this value.

Hysteresis is the lagging effect on a magnetic device when the
magnetizing force is applied. It also represents the delaying effect of
the collapse of the magnetic field in the device when the magnetizing
force is removed.

Degaussing is the act of neutralizing any magnetic field present in
the device. Degaussing coils are used commercially to accomplish this.
The coil is connected to an alternating source of electrical energy.
This alternating force balances the magnetic charges on the device.
Once these charges are balanced, they cancel each other.

A wire carrying an electric current has a magnetic field around it.

FORMULAS

Magnetic field strength:

$$H = \frac{\mathfrak{F}}{l}, \text{ in amperes per meter}$$

Magnetomotive force:

$$\mathfrak{F} = AT, \text{ in amperes}$$

PROBLEMS

8-1. A coil of wire has 750 turns. Total circuit current is 0.350 A. Find the ampere-turns.

8-2. A coil of wire has 0.065 A of current. It consists of 450 turns of wire. Find the ampere-turns.

8-3. A coil of wire has 650 turns. The current flowing through it is 0.150 A. Find the magnetomotive force.

8-4. A coil is wound with 3,000 turns of wire. A current of 0.250 A flows through it. Find the magnetomotive force.

8-5. A coil has a magnetomotive force of 650. It has 4,000 turns of wire. What electrical current flows in the coil?

8-6. A coil has a magnetomotive force of 1,200. It has 3,200 turns of wire. What current produces this force?

8-7. Find the magnetic field strength of a coil 50 cm long whose MMF is 650.

8-8. A coil is 150 cm long. It has 300 turns of wire. It has a current of 7.15 A. Find the magnetomotive force and the strength of the magnetic field.

8-9. A coil is 15 cm long. Its current is 0.06 A. The magnetomotive force is 12 A. Find the number of turns of wire in the coil and the magnetic field strength.

8-10. A coil has 12,000 turns of wire. Its current flow is 0.010 A. It is 400 cm long. Find the magnetomotive force and the strength of the magnetic field created.

QUESTIONS

8-1. Magnetic lines of force

    a. move across the width of a magnet

    b. run from one pole to another on a magnet

    c. always touch each other

    d. travel in straight lines

8-2. Magnetic lines of force are strongest

    a. at a magnet's poles

    b. midway between magnetic poles

    c. 70.0% away from each pole

    d. at the north magnetic pole

8-3. The north magnetic poles of two magnets

    a. attract each other

    b. have no effect on each other

    c. repel each other

    d. produce a third magnetic field

8-4. The number of lines of force produced by a magnet per square inch is called

    a. permeability

    b. flux

    c. flux density

    d. core strength

8-5. The ability of a material to become a magnet is called

    a. retentivity

    b. reluctance

    c. permeability

    d. flux strength

8-6. An electromagnet's strength is measured in units of

    a. ampere-turns

    b. permeability

    c. flux density

    d. retentivity

8-7. The lag in the reduction of a magnetic field's strength is called

    a. flux retention

    b. permeability

    c. ampere strength

    d. hysteresis

8-8. The process of neutralizing a magnetic field is called

    a. flux reduction

    b. degaussing

    c. reverse magnetism

    d. none of the above

8-9. Magnetic strength of a permanent magnet is limited by its

   a. size

   b. shape

   c. composition

   d. all of the above

8-10. Two bar magnets are placed end to end and have their opposite poles touching. The overall strength of the combination is ____ that of one magnet.

   a. stronger than

   b. weaker than

   c. the same as

   d. has no effect on

9

Electromagnetic Inductance

The relationship between magnetism and electricity is very close. Both of these phenomena are useful in an industrialized society. Production of the tremendous amounts of electrical energy required for today's world would be either impossible or extremely expensive if done without the use of magnetism.

This chapter explains how magnetic principles are related to electrical principles. It also explains how magnetism is used in electrical components and devices. Later chapters show how the units described here are used in business, home, and industry.

ELECTROMAGNETISM

The magnetic strength of a permanent magnet is limited by its size and the cost of producing this kind of magnet. A permanent magnet has many limitations. For example, one cannot turn off a permanent magnet. Also, the strength of the magnetic field cannot be varied unless the magnet is moved from one position to another position. These limitations, and others, are overcome when an electromagnet is used.

An electrical current moving through a conductive material, such as wire, will produce magnetic lines of force. These lines of force surround the conductor. They also form at right angles to the conductor. This is shown in Figure 9-1, which shows a small section of a piece of wire. An electrical current is flowing through the wire. This creates the magnetic lines of force. The strength of the magnetic field pro-

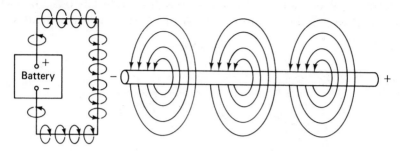

Figure 9-1. Electromagnetic fields are formed around any current-carrying conductor. They form at right angles to the conductor and completely surround it.

duced in this manner is directly proportional to the amount of electric current flowing through the wire. The magnetic field size increases as more current flows. A reduction in current makes the magnetic field smaller. It is very difficult to show, but picture the magnetic field completely surrounding the conducting wire. This magnetic field is very strong close to the surface of the wire. It diminishes in strength as the distance from the wire increases.

Magnetic lines of force have direction. Electromagnetic lines of force also have direction. There is a rule that demonstrates the lines of force surrounding a conductor. This is called the left-hand rule. It is illustrated in Figure 9-2. The thumb is pointed in the direction of

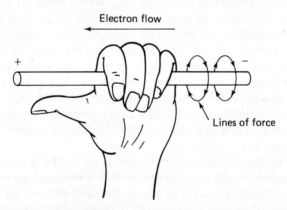

Figure 9-2. The left-hand rule for determining current flow in a conductor.

the electron flow in the wire. This is toward the positive terminal of the power source. The fingers are curved around the conductor. They show the direction of the flux lines surrounding the wire. These lines of force are continuous. They do not have any polarity. They do, however, provide polarity to other magnetic materials brought near them. Figure 9-3 shows this effect. A piece of soft iron is held near

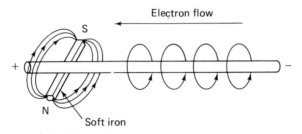

Figure 9-3. Electromagnetic forces from the conductor will produce a magnetic field in the soft iron rod.

the current-carrying conductor. The soft iron forms a magnet. The magnet has poles. The poles remain, as does the magnet, until the electrical current is stopped. When this occurs, the soft iron no longer is a magnet. One may wish to try this in order to "see" how it works. Use a large nail or other soft iron material. Changing the distance between the nail and the wire will provide a change in the magnetic field in the nail. As the nail is moved away from the current-carrying wire, its magnetic properties are reduced. This relates to the strength of the magnetic field around the wire and the reduction of the magnetic field as its distance from the wire increases.

Another experiment is to change the relationship of the nail and the wire. They have been at right angles to each other. Rotate the nail so that it is parallel to the wire. The strength of the magnetic field in the nail should be reduced. This shows that the lines of force from the wire have most effect when the iron is at right angles to the wire. Each line of force surrounds the wire and is at right angles to the wire. The maximum effect of the lines of force on the nail occurs when they are at right angles to each other, because a greater number of lines of force cut across the nail. This principle applies in all electromagnetic devices.

Reversal of the direction of current flow in the wire will reverse the poles created in the soft iron piece being magnetized by the electrical current. This is done by reversing the connections on the power source. All the other events relating to magnetization and the strength of the magnetic field remain the same. The only change occurs in the polarity of the magnet created by this magnetic field.

Very often two or more individual wires are placed next to each other. Sometimes these wires are manufactured as a "pair". The most common examples of this are the power cords found on home appliances. Each individual wire will show magnetic effects when the electric current flows in the wire. The magnetic effects interact with each other. This is shown in Figure 9-4. When each wire has an electric cur-

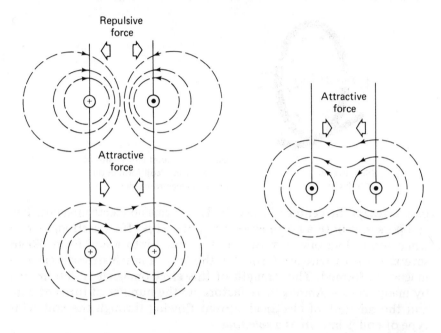

Figure 9-4. Each wire of a multiconductor cord has its own magnetic field formed by the current flow. This action produces either an attraction or a repulsion of other conductors in the cord.

rent flowing in the same direction, one wire's magnetic field will oppose the other wire's magnetic field. Since like charges repel each other, the wires tend to try to move apart. The magnetic fields created in this situation are weak. This is particularly true when the wires are straight as opposed to being wound up on some sort of form. When one wire's electrical current flow is opposite that of the other wire, the two fields attract each other. Magnetic lines are strongest in the area between each of the wires. This is true regardless of the direction of the electron flow through each wire.

THE INDUCTIVE COIL

The electromagnetic theories presented to this point in this chapter deal with the relationship of the field when an electrical current flows through a straight wire. In many cases the wires are used in another form. This form is often circular in shape. Using wires wound in the form of a coil gives a different shape to the magnetic field. A

look at Figure 9-5 will help in understanding how this works. When
the wire is wound in the form of a coil, all the magnetic lines of flux

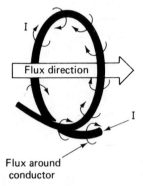

Flux around
conductor

Figure 9-5. Conductors are often wound in a coil form in order
to form an electromagnet.

pass through the center of the coil. This forms an electromagnet. The
coil possesses both a north and south pole. This effect is strengthened
when more than one turn of wire is formed into a coil form. When
several turns of wire are formed in this way, a much stronger electromagnet is formed. The strength of this electromagnet is determined
by many factors. Among these factors are the number of turns of wire
and the amount of electrical current flowing through the coil. This
type of coil is known as a *solenoid*.

The solenoid acts exactly like an electromagnet. A soft iron core
is used to strengthen the magnetic field of an electromagnet. This
also is true for the solenoid. Using a soft iron core material tends to
give the magnetic lines of force inside the coil winding less opposition to movement. This reduction of the reluctance of the magnetic
properties produces a stronger magnetic field when compared to the
magnetic field produced with no coil core.

Keep in mind that the only time the coil of wire acts as a solenoid
is when an electric current is flowing through the wire. In other words,
the solenoid must be connected to a source of electrical power. An
elecromagnet is produced when the power is on. Turning off, or disconnecting, the source of power also turns off the solenoid electromagnet. This principle is important. It applies to most electromagnetic
circuits. Now we have the capability of creating an electromagnetic
field when it is needed. This allows us to use this form of magnetism
to perform work.

ELECTROMAGNETIC INDUCTION

Very few wires used in electrical devices are perfectly straight.
They seldom are exactly parallel to each other. The electromagnetic

flux fields produced when an electrical current flows through the wire will interact with either other fields or other wires. They also react with another section of the same wire. Let us explore and discuss these actions.

When a magnetic field cuts across a conductor, the magnetic lines of force produce a voltage on the wire. This is illustrated in Figure 9-6.

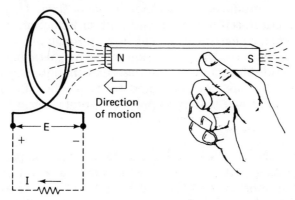

Figure 9-6. Moving a magnetic field past a conductor will induce a voltage on the conductor.

This voltage, or electromotive force (EMF), may be produced in one of two ways. Moving the magnetic field across the wires is one way. The other way is to move the wires past the magnetic field. In either way the lines of force from the magnet are cutting across the wire. Doing either of these will produce an EMF on the wire.

An adaptation of this method is shown in Figure 9-7. Here the

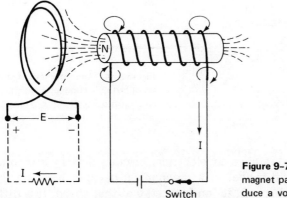

Figure 9-7. Moving an electromagnet past a conductor will produce a voltage on the conductor.

magnetic field created in the solenoid coil is changing. The change occurs when the electrical energy that forms the magnetic field is either turned on or turned off. This illustrates how a changing magnetic field is able to produce an EMF on a wire.

These processes are called *electromagnetic induction.* An EMF is *induced* from the first circuit on to the second circuit. There is no direct electrical connection between these circuits. The EMF is induced by means of the lines of force. To be effective, the lines of force that induce the EMF have to be moving. No induction occurs if the lines of force and the coil are stationary. The induced EMF gives the wire a polarity. This produces an electrical current flow. There are two types of induction, self-induction and mutual induction. Self-induction occurs when an electrical current flows through one coil of wire. Self-induction describes the interaction of the magnetic lines of force in the same coil. Mutual induction describes the interaction of two wires or coils that are not connected to each other. Both of these properties are used successfully to produce useful electrical energy.

Self-Induction

Self-induction occurs when the lines of force from one part of a coil cut across another part of the same wire in the same coil. If the lines of force are in motion, an EMF will be induced. The lines of force are produced by the electrical current flowing in the coil. The electric current has a specific polarity. The induced EMF is of opposite polarity when the force field is increasing in size. The induced EMF is of the same polarity as the applied current when the force field is collapsing. This leads to a basic rule regarding induced EMF. The rule states that the induced EMF and its resulting current develop a flux that tends to oppose the forces which induced the current. This rule applies in both forms of inductance. It is called Lenz's law. Figure 9-8 illustrates this law.

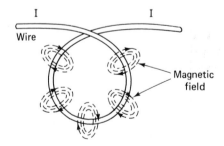

Figure 9-8. The magnetic field produced by a conductor which is wound into a coil will produce a voltage on other parts of the same conductor.

In one sense, inductance in electrical circuits acts in a manner similar to that of a flywheel in a mechanical device. Once the flywheel is at rotational speed, it tends to stay at that speed. It is difficult for changes in the load to vary the flywheel speed. Energy is stored in the flywheel as it rotates. This tends to oppose any changes in its speed. There is an electrical device that acts in a similar manner.

It is a coil of wire. The coil of wire is called an inductor. The flywheel effect in the electrical circuit is called inductance. Once a specific current flow is established in the circuit, the inductance tends to maintain that current flow. It reacts to changes just as the flywheel reacts. The energy stored in the inductor tends to maintain that specific level of current flow. A drop in current flow makes the flywheel give up some of its energy. A rise in current flow is absorbed by the flywheel. The result is a constant current flow in the circuit. This continues until the energy in the flywheel is exhausted.

The magnetic flux created as current flows in the coil of wire produces magnetic fields. They will cut across the wire in several places. An EMF is produced when this occurs. The EMF has an opposite polarity to that of the EMF producing the magnetic field. The induced EMF tends to oppose any change in the current flowing through the wire. Since the induced EMF is of opposite polarity, it is called *counter EMF* or *back EMF*. Counter EMF is produced only when the flux field is changing in size. It does not occur when the field is stationary. This is illustrated in Figure 9-9, which shows an electrical circuit con-

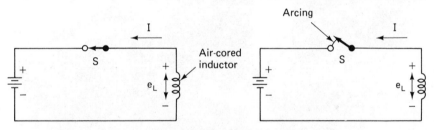

Figure 9-9. The energy stored in an inductor will produce a current flow and arcing across an open circuit when the source of energy is removed.

taining a source of electricity, an inductor, and a switch. The polarity of the source is shown. When the switch is in the on position, as is shown on the left side, current flows from the negative connection of the source, through the coil and switch, and then back to the positive connection on the source. The EMF induced in the coil produces an opposition to the flow of current. This tends to reduce overall current flow in the circuit. This type of circuit opposes the rise of current. The opposition continues as long as the current in the circuit increases. Once the current reaches a level where it stabilizes, the lines of force produced by current flow no longer grow. When the switch is in the off, or open, position, the current no longer flows in the curcuit. The lines of force created by the current flow start to collapse. They now are moving in the opposite direction. When they move in the opposite direction, they produce an EMF that supports the polarity of the applied voltage. They now aid the original current flow of the circuit. The EMF produced in this manner attempts to maintain circuit current even though there is no applied voltage at this time.

The amount of EMF produced in this manner is directly dependent on the amount of inductance of the coil. The coil's inductance is measured in units of the henry (H). The letter symbol for inductance is L. One henry of inductance is created when an EMF of 1 volt is induced by a current changing at the rate of 1 ampere per second. The terms used to describe a coil are usually *inductor* or *choke*. The term choke describes the reaction to change of current in a circuit containing inductance. The inductor tends to "choke off" any changes in circuit current, hence its name. Pictures illustrating typical inductors are illustrated in Figure 9-10.

Figure 9-10.   Common inductors used in the electrical industry.

Mutual Inductance

Another form of inductance is mutual inductance. Mutual inductance describes the action of magnetic flux from one wire or coil as it crosses a second wire or coil. The magnetic flux created by a changing electrical current may be used to develop an EMF in another circuit. The second circuit is electrically insulated from the first circuit. There are no direct connections between the two circuits. This is illustrated in Figure 9-11. Lines of force from the left coil cut across the right coil. This induces EMF in the second coil. This type of circuit is called a *transformer* circuit. The coil of wire that is connected to a power

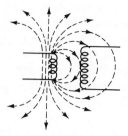

Figure 9-11. Transformer action occurs when lines of force from one coil cut across wires in another coil.

source is called the *primary winding*. Coils of wire that develop the induced EMF are called *secondary* windings or coils. Some transformers use a core material to give direction to the lines of force. A variety of core shapes and windings are also used. Specific use will depend upon application. Mutual inductance, like self-inductance, is measured in henries. The laws relating to the polarity of the induced EMF also apply in mutual inductance circuits.

When the flux field increases in size, the EMF produced is of opposite polarity to the EMF that produces the flux field. When the flux field is decreasing, or collapsing, the EMF created is of the same polarity as the EMF that produced the flux field. As previously stated, the EMF is created in the secondary windings only when the flux field is changing. There is no EMF created when the field is stationary.

The law for mutual inductance is very similar to the law for self-inductance. Two coils have a mutual inductance of 1 henry when an EMF of 1 volt is induced in one coil by a current changing at the rate of 1 ampere per second in the other coil.

ENERGY TRANSFER

The amount of energy transferred from the primary to the secondary of any device with mutual coupling depends upon several factors. These factors all relate to the magnetic flux field created in the primary windings of the device. First, most devices using the principles of mutual inductance are called transformers. The magnetic flux field created in the primary of the transformer depends upon wire size, spacing between turns of the wire, the value of the EMF producing the flux field, and the total number of turns of wire in the primary. The current-carrying capacity of any conductor is directly related to the physical size of the wire. The capability to carry larger amounts of current increases with the size of the wire. Spacing between the turns of wire of the primary is related to the magnitude of the flux field. Winding spacing, or length, is the reciprocal of the number of lines of force in a specific area. Close windings will produce a maximum flux density. All wires have an opposition to the electric current flow in the wire. This opposition does not change. A stronger EMF will force more current through a wire. A weaker EMF forces less current through the same piece of wire. Another factor is the total number of turns of wire on the coil. As the number of turns is increased, the total magnetic flux field increases. These factors all have an effect on the size of the magnetic flux field created in the primary of any transformer.

Reviewing the factors affecting the creation of an EMF in the secondary of a transformer shows that all the previously described factors also influence the EMF created in the secondary. Other factors also bear on the transfer of energy in a transformer. These include the spacing between wires in the primary and secondary windings, the manner in which the wires are wound, and the type and shape of the core material used in the transformer. Lines of force in any flux field are strongest nearest to whatever produces the field. Placing primary and secondary winding wires next to each other on the transformer provides a maximum of energy transfer. This also relates to the manner in which the coils are wound. The type and shape of the core material relates to energy transfer, also. Shape is a factor left primarily to designers. Shape and the kind of core material are factors also related to the operating frequency of the transformer. These factors will be discussed in detail in Chapter 13.

Transformers are important in electrical applications. Use of the transformer provides a variety of EMF levels. Transformers also have given us the ability to transport electrical energy over great distances.

MOTOR ACTION

Principles related to magnetic flux fields and their interaction may be used to produce rotating machinery. One of the most common rotating machines known today is the electric motor. The application of the principles of magnetic attraction and repulsion are used in explaining how the electric motor works. The lines of force in any magnetic field have a specific direction. These are similar to the lines of force from an electromagnetic field. The left-hand rule for solenoids describes this factor. If one magnet is placed between the force field of another magnet, as shown in Figure 9-12, the two fields interact. If one magnet is at right angles to the other magnet, maximum interaction between the two fields occurs. The magnetic field on one side is reinforced. The magnetic field on the opposite side is weakened. This provides motion to the first magnet. The direction of the

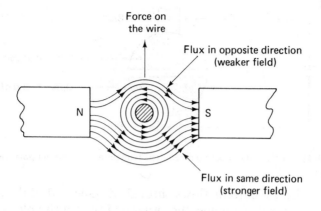

Force on
the wire

Flux in opposite direction
(weaker field)

N   S

Flux in same direction
(stronger field)

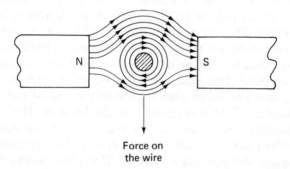

N   S

Force on
the wire

Figure 9-12. Motor action produces a rotary motion due to the interaction of two magnetic fields.

force depends upon the relationship of the two magnets. Reversal of either set of magnetic poles will reverse the direction of the force.

The amount of force created in this manner is limited to the strength of each of the two magnets. Either one of these may be made stronger by substituting an electromagnet. Rotary motion is created when the magnet in the center is placed on an axle. This rotating magnet is called an *armature*. The fixed magnet surrounding the armature is called the *field*. Electrical energy is connected to the armature by a device called a *commutator*. Carbon brushes ride against the commutator. The brushes provide contact between the commutator and external wiring. This is illustrated in Figure 9-13.

The commutator is required to provide a continuous rotary motion to the motor. Without the commutator, the shaft on which the armature is placed would only turn 180 degrees and then stop. This is because the armature has a specific polarity, which is developed

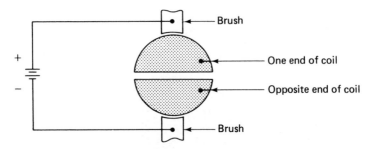

Figure 9-13. Commutator action is required to maintain the correct polarity on the armature.

when the electric current flows through it. One end of the armature becomes the north pole and the other end the south pole. When this situation occurs, the armature would rotate once until its north pole and the field's south pole were near each other. At this point, all rotary motion stops. The opposite magnetic fields are aligned. Continued circular motion occurs when one set of magnetic fields is reversed. This is accomplished by the use of a commutator. The commutator is split into equal parts. One side is connected to one end of the armature winding, and the other end to the other end of the armature winding. A constant EMF is connected to the brushes. The commutator is mechanically connected to the armature. It is placed on the same shaft as the rotating armature. Rotation of the commutator and armature changes the polarity of the EMF applied to the armature. This change is designed to occur so that the rotation of the armature is continuous.

The type of electric motor described here has little practical value. The only time the motor is developing any torque is during that portion of its rotation when the armature poles are closest to the field poles. It has a tendency to coast during the balance of its cycle. This is overcome in a practical motor by adding more armature winding segments.

GENERATOR ACTION

The electric motor previously described creates rotary motion from a source of electrical energy. It can be stated that the motor changes electrical energy into mechanical energy. Work is accomplished when this occurs. Mechanical energy may also be changed into electrical energy. This is accomplished by means of an electric generator. This is illustrated in Figure 9-14. It has been stated that an EMF will be induced on a wire when the force field is moving across the wire. This is the principle of generator action shown in the illus-

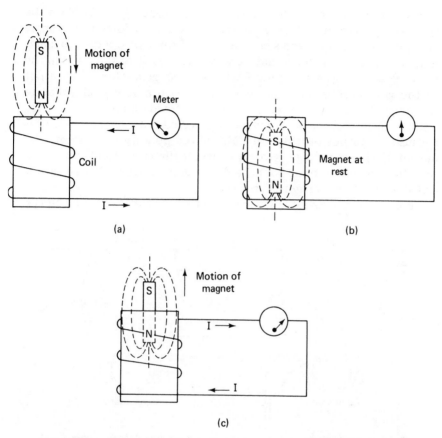

Figure 9-14. An electric generator is formed when one magnetic field is moved past or through a coil of wire. The polarity of the induced voltage depends upon the direction of movement of the magnetic field.

tration. Moving the magnetic field into a coil of wire creates an EMF on the coil. It gives direction to the flow of electron current in the coil as indicated on the electric meter. Stopping the movement of the magnet also stops the development of an EMF and the resulting electron flow. Removing the magnet from the coil produces an EMF of opposite polarity. This makes the electron current flow in the opposite direction. All three of these conditions are illustrated by use of an electric meter in Figure 9-14.

This same principle is used in the generation of electrical energy. A machine very similar to the electric motor is used. The simplest generator has a permanent magnet as its field. The armature is rotated through the field by means of a device connected to the armature shaft. This may be a water wheel or a steam or gasoline engine. Rotation of the armature in the magnetic field produces an EMF on the wires of the armature. This EMF is directly proportional to the angle at which the armature wires cut the lines of force of the field magnet. It is strongest when the armature wires are closest to the poles in the field. This produces a varying EMF from the generator. The output of the generator is made more constant by use of several armature windings. A split commutator is used to provide a constant polarity output EMF. This type of generator is called a dc generator. This is because its output provides an EMF whose polarity is constant. The output EMF, or voltage, gives a constant direction to electron current flow in any circuit connected to the generator. This voltage is called dc, or direct current, for this reason.

Figure 9-15 illustrates the characteristics of this output voltage.

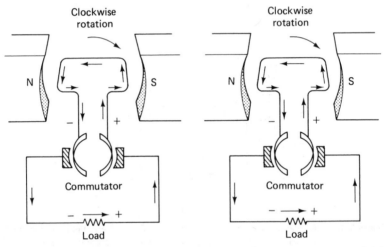

Figure 9-15. The DC generator. The generator requires a commutator to maintain a constant polarity output voltage.

The horizontal line in the graph represents time. The armature of the generator takes a period of time to complete one rotation of 360 degrees. This provides a dimension of time to the production of the voltage from the generator. The vertical line represents the magnitude and the polarity of the voltage produced by the generator. In this situation the generated voltage starts at zero at the first instant of time. It then rises to a maximum value as the armature approaches the pole of the field. As the armature moves away from the field pole, the voltage produced reduces to zero. At this point the commutator switches the wires between the armature windings and the outside connection. Now the other end of the armature swings past the field pole. The switching of the wiring by commutator action maintains a constant polarity output voltage.

There is a relation between the laws for electromagnetic induction and the voltage created by generator action. Increasing the rate at which the wires in the armature cross the magnetic flux lines will increase the magnitude of the output voltage. More wires on the armature will also increase the induced voltage. Adding more segments to the armature makes it a much more efficient device.

A second type of generator is the ac generator. The basic construction for this type of generator is the same as that for the dc generator; the only major difference is that devices called *slip rings* are used in place of the commutator. Their purpose is also to connect the rotating armature to the devices requiring this type of voltage. The slip rings are circular bands of copper. They are fastened onto the shaft holding the armature. Brushes made of graphite ride against the slip rings as they rotate. This is illustrated in Figure 9-16.

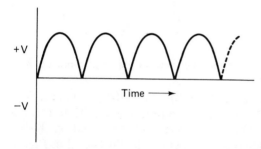

Figure 9-16. The characteristics of a DC voltage. This voltage rises from zero to a maximum value and then drops to zero as the armature passes the pole of the armature magnet. The commutator action maintains the direction of current flow in the circuit.

Reversing the polarity of the magnetic field crossed by a wire will reverse the polarity of the induced EMF. This in turn reverses the direction of electron current flow in the generator. Each pole of the armature crosses magnetic flux fields, which produce first an EMF of one polarity and then an EMF of the opposite polarity. Thus, the induced EMF alternates between positive and negative polarity. This action produces an alternating electrical current in the devices connected to this generator. This type of generator is called an ac generator. It is often referred to as an *alternator*.

Figure 9-17 illustrates the output EMF action of this type of generator. Use the horizontal axis to represent time. The vertical axis of the graph will be used to represent the amount and polarity of the induced voltage. Only one end of the armature winding is used as the point of reference for this diagram. It turns through 360 degrees to complete one cycle or rotation.

Assume that the armature is started when its pole is equidistant from the field poles of the generator. The EMF induced at this time is minimal. Call it zero EMF. Rotating the armature brings it close to the north pole, back to a point between the poles, to the south and back to its starting point. The induced EMF follows this very closely.

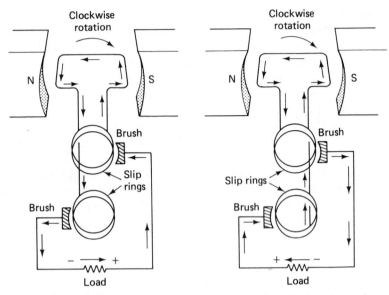

Figure 9-17. The AC generator. This generator requires slip rings to produce a varying polarity voltage to its load.

The EMF starts at zero, rises to a maximum value, drops to zero, rises again to a maximum value of the opposite polarity, and then returns to zero at its starting point. The shape of the wave created by the action of the ac generator is called a sine wave. It is illustrated in Figure 9-18. Its rate of repetition is called its *frequency*. These terms and others related to ac generation are discussed in detail in Chapter 10.

SUMMARY

There is a very close relationship between magnetism and electricity. A permanent magnet's strength is limited by its size and the type of material from which it is made. These limitations are overcome by the use of an electromagnet. Both types of magnets have similar properties.

An electric current moving through a wire will create a magnetic field around the wire. The strength of the magnetic field created in this manner is directly proportional to the strength of the current. All magnetic lines of force have direction. There is a definite polarity developed by these lines of force. Both north and south poles are developed.

A wire that is wound in the form of a coil will form an electromagnet. The electromagnet also has poles and lines of force. The strength of the lines of force is found at each pole of the electromag-

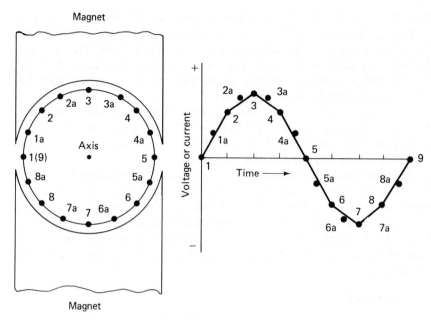

Figure 9–18. The characteristics of an AC voltage. This voltage also rises from zero to a maximum value. It alternates its polarity owing to the action of the slip rings.

net. This coil is called a solenoid. An iron core material in the solenoid will give more direction to the lines of force and make a very strong magnet.

Lines of force from a magnetic field can cut across a wire. When this occurs, a voltage is created on the wire by a process called electromagnetic induction. This occurs when either the field or the wire is in motion. It does not happen when both are stationary. When induction occurs in one coil's wires, it is called self-induction. When this process occurs between two wires that are not connected to each other, it is called mutual induction.

Inductance in an electric circuit is like a flywheel in a mechanical device. Inductance reacts to circuit current changes by attempting to hold the current at a specific level. Rises in current are absorbed by the circuit inductance. Drops in current are replenished by the inductive energy in the circuit.

The EMF that is induced on a wire has a definite polarity. When EMF is induced by a rising current, it is of an opposite polarity to that of the applied voltage. EMF induced by a falling current is of the same polarity as the applied voltage.

The unit of inductance is the henry (H). Inductance is represented by the letter L. Inductors are often called chokes. This is because they tend to choke off, or oppose, any circuit current changes.

Energy is transferred from one circuit to another by inductive action. The device used for this purpose is a transformer. The amount of energy transferred depends upon the strength of the magnetic field, the physical relationship of the wires, the size of the wires, and the core material used in the transformer.

Interaction between two force fields will produce rotary motion. This is called motor action. One field is called the armature. It rotates on an axle. The other is called a field winding. Wires connect to the rotating armature by means of a commutator.

A similar principle is used in the generator. An armature is rotated by external means. It moves around in a magnetic field. The armature wires have an EMF induced on them. This device is called a generator. When the generator output is a constant voltage and polarity, it is called direct current, or dc. When the output is constantly varying, it is called alternating current, or ac. An ac generator is often called an alternator.

QUESTIONS

9-1. Magnetic flux is created around a wire when

    a. a current flows in the wire

    b. there is a voltage near the wire

    c. power is beamed at the wire

    d. all of the above

9-2. The process of producing magnetic properties in a wire or conductor is called

    a. reduction

    b. magnetism

    c. induction

    d. opposition

9-3. Production of an EMF in one wire from another wire or circuit occurs when

    a. the current changes in the first wire

    b. the current is constant in the first wire

    c. there is no current flow

    d. none of the above

9-4. The EMF produced in a secondary circuit is called

    a. counter EMF

    b. forward EMF

    c. applied voltage

    d. reverse current

9-5. The EMF produced in a secondary circuit has ____ polarity to that in the primary circuit

    a. the opposite

    b. the same

    c. no effect on the

    d. equal

9-6. A device used to transfer energy from one circuit to another is called

    a. a solenoid

    b. a magnet

    c. a transformer

    d. a reverse solenoid

9-7. Electrical energy used to create rotary action is called

    a. generator

    b. motor

    c. transformer

    d. rotor machine

9-8. Rotary energy used to create electrical energy is called

    a. generator

    b. motor

    c. transformer

    d. rotor machine

9-9. The device used to connect the motor armature to the load is called the

    a. rotary convertor

    b. commutator

    c. armature extractor

    d. take off connection

9-10. Generator action is strongest when the armature winding is ____ to the poles of the field winding

    a. farthest

    b. at mid-point

    c. 20 degrees away

    d. closest

10

Generation of Voltage

Alternating current sources are used almost exclusively to provide operating power for home and industry. This is because, under normal conditions, ac is more economical to transport over long distances. This permits distances of many miles between the user and the producing station. Some inroads have been made in recent years with new technology and dc sources. These are in experimental development stages at the present time. They are not readily available for large commercial use.

Alternating current has many uses and forms. Basically, it is a varying level voltage or current. Its rate of variation may change, or it may be constant. The amplitude of the ac may change, also. Often the actual shape of the ac wave changes. All these are characteristics of the ac wave. It is a good idea to distinguish between two of the major categories of ac. The largest use for ac is to provide electrical power that will operate machinery, lights, and a large variety of electronic devices such as tape players, radios, and television sets. The second most common use for a varying ac is for the information processed by computers, radios, and television sets. This second use is called signal ac. In most instances of information processing, both of these ac voltages are present and used. If confusion has set in at this point, the following may help to organize the two types of ac.

1. **Application.** Power for operation is simply called *power*. This is used to provide operating voltage and current for the device. The device may be a lamp, an electric motor, a computer, or a television set. Use of ac as an information-carrying device is referred to as *signal*

ac. The signal may be broadcast from a radio station, found on a recording tape, or be placed on a telephone line and sent from one station to another station.

2. **Frequency.** Basically, if only one frequency is used, then the
ac is called power. Most of the frequencies used for commercial power
in the world are between 50 and 60 repetitions per second. These are
power frequencies and are used to provide operating power to electrical devices. If the frequency is much higher than those used for
power, or if it varies across a large range, it is probably a signal. Another characteristic of the ac signal is that it may contain several frequencies at the same time. As is the case with most things, there are
exceptions to these classifications. If you are able to classify ac by
these two means, this will help you to understand what electric utilization is all about.

The second major classification for voltage is dc. Direct current
flows in only one direction, hence its name. The uses of dc as a power
source for major electrical devices are limited. More often dc is used
in electronic devices as a power source. These devices require smaller
amounts of power. Some devices use batteries as a dc source. Others
change ac into dc in order to provide the proper operating conditions.

This chapter presents material covering the production, distribu
tion, and form of the various voltages used in electrical devices. Sig
nals and their composition are also covered. These two types of elec
trical waves are basic to most electric products.

ALTERNATING CURRENT

Alternating current, or ac, is considered to be a constantly varying value. Alternating current may be used to describe both voltage
and current. Most of the ac produced in the world is used to power
electrical equipment. The shape of the waves is sinusoidal. Several
characteristics can be used to describe any electrical wave. These include shape, frequency, timing, and amplitude. These terms as applied
to ac waves are discussed in this section.

Transportation of the AC Wave

The first question that often presents itself is, "Why do we use ac?"
This can best be answered by stating that ac is easier to transport
from the generating plant to the consumer than any other form of
electric power. A characteristic of electricity is resistance, which opposes the flow of electrical current. In any electrical circuit there is
some resistance in the wires that connect the consuming device, or

load, to the source. If this wire resistance is high, the voltage at the load end of the circuit is lower than the voltage produced by the generator. This is illustrated in Figure 10-1. The resistance in the wires

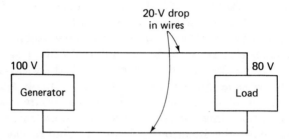

Figure 10-1. A drop in voltage due to the resistance of the wires results in lower voltages at the load.

connecting the source to the load causes a 20-volt drop at the load. This drop in voltage reduces the efficiency of the load. This type of voltage drop happens in any conductor. It may be reduced to a minimum value by increasing the size of the conductors. The length of the wires between the source and the load is also a factor to consider. Long wires introduce larger amounts of resistance into the system. It is easier to overcome these problems by the use of ac than it is when using dc.

Electric power is a product of voltage and current. Electric power provides the energy to perform work with electricity. The formula $P = E \times I$ is the basic power formula. P (power) = E (voltage) $\times I$ (current) shows that increasing voltage and dropping the current will produce the same amount of power. For example

$$1,000 = 1,000 \times 1$$

or

$$1,000 = 500 \times 2$$

or

$$1,000 = 2 \times 500$$

The power remains the same as long as the ratio of the voltage and current remains the same. This means that either voltage or current may be increased and the total power will remain constant. This is used in the transportation of electrical energy.

Power producers wish to generate power and deliver it with a minimum of loss. Losses represent reduced efficiency as well as financial drain. This loss is most apparent in the conductors carrying current between source and load in the system. It could be overcome by using very large conductors. This, however, is expensive, for the cost of large conductors is relatively high. The cost of the towers, or poles, to hold very heavy conductors is also relatively high. These factors are overcome by changing the value of the voltage in relation to the current in the system. Most often the voltage is increased to a very high value. This is done by means of a transformer at the source. Transformers are very efficient devices. They vary from 80% to 98% efficiency. The ratio of numbers of turns of wire between primary and secondary windings allows voltages to be *stepped up*, or increased. Reversing this permits voltage to be reduced, or *stepped down*. The following is an example of how this is used by the electrical power industry.

The electric power industry generates about 18,000 volts at the source. The current produced at our model power plant is 20 amperes. Using the power formula, $P = E \times I$, the power produced is 360,000 watts. Assume that the resistance of a pair of wires carrying this power to a local distribution point is 250 units or ohms. The voltage loss is figured by use of the formula $E = I \times R$, where E equals the voltage loss, I represents the current, and R is the resistance or ohms. The voltage drop, or loss, would be 250×20, or 5,000 volts. The voltage available at the load end of the circuit would be the difference between the generated voltage and the drop voltage, or 13,000 volts. If the current remained at 20 amperes, there would be a power loss. The power at the load is $13,000 \times 20$, or 260,000 watts. The efficiency of this system is about 72%. This is too low an efficiency for power companies. The losses are too large to be economically sound.

There is a very easy way to overcome this high rate of loss. This is accomplished by the use of a transformer at the source to step up the voltage. The voltage is increased through the very efficient transformer to about 345,000 volts. Increasing the voltage and maintaining the original amount of power will give a reduction in the current flow. The power formula $P = E \times I$ gives us this information. If the power of 360,000 watts is produced at 345,000 volts, then the current is 1.043 amperes (345,000/360,000 = 1.043). Use these values in the same system with 250 ohms of line resistance. The result is a voltage drop of 260 volts instead of 5,000 volts. There is a higher level of voltage available at the load. The efficiency of this system is about 96%. This is a much more practical approach than when a lower voltage system is used.

The system described here is a model. Power-producing companies use this method to increase the efficiency of their systems. The values of current and resistance used are not necessarily true values. They are given to offer a reasonable example of why higher voltages are used when power is transported. There are other inefficiencies in the system. These will cause additional power losses. Each system is engineered in a way that minimizes these losses. A typical distribution system is illustrated in Figure 10-2. This shows both the step-up sys-

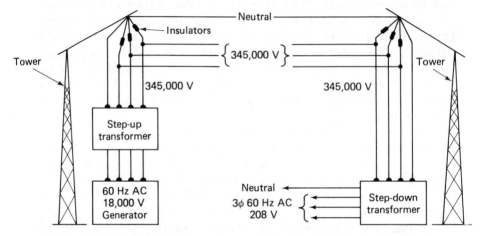

Figure 10-2.　Power distribution systems use high voltages and low currents in order to transmit a maximum amount of power with minimal losses due to resistance of the wires.

tem at the generating plant and the step-down system used at the load distribution end of the systems. This type of network is very similar to those used by local power companies when supplying power to homes.

Sine Wave

The rotation of an ac generator's armature through a complete revolution will produce a varying voltage. This voltage varies both in the amount of voltage produced and in the polarity of the voltage. If this voltage was measured at each instant of time during one complete revolution of the armature, and the measured values were plotted on a chart, the result would be what is known as a sine wave. This is illustrated in Figure 10-3. This shows the position of the armature as it rotates and the relative voltage produced at each major point of reference during one revolution of the armature coil. This shape is the form of the ac wave. It is called a sine wave. The form of the ac wave is sinusoidal.

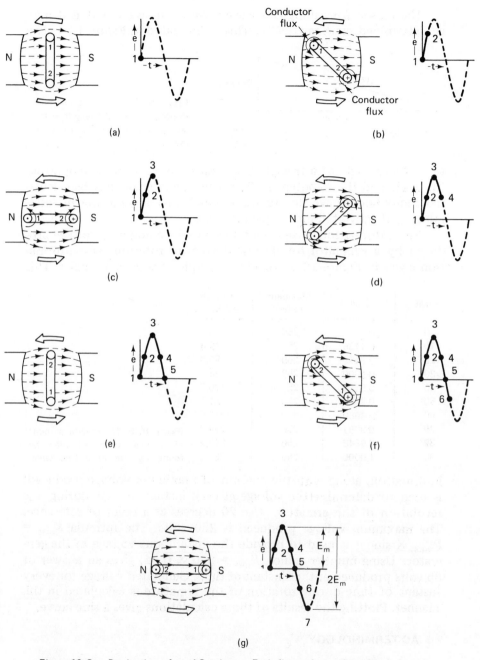

Figure 10-3. Production of an AC voltage. Each figure shows the voltage created by the armature moving past a polar magnet in the generator.

The reason for calling this wave form a sine wave is that it may be determined mathematically. This is illustrated in Figure 10-4. The

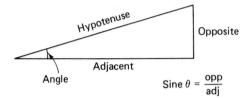

Sine $\theta = \dfrac{\text{opp}}{\text{adj}}$

Figure 10-4. Use of a sine relationship to find the length of the leg opposite to the angle of movement.

sine of any angle of a triangle is the ratio between the side opposite the angle and the hypotenuse. The sine of any angle may be calculated. Many books have sine tables in which this angle has been determined and converted into a specific value.

This information is used when figuring the amount of voltage produced by a generator for specific degrees of rotation. The information given in Figure 10-5 provides a sample of some sine angles. This

Angle	Sine	Maximum voltage	Voltage generated
0°	0	250	0
10°	0.1736	250	43.4
20°	0.3420	250	85.5
30°	0.5000	250	125
40°	0.6828	250	170.7
50°	0.7660	250	191.5
60°	0.8660	250	216.5
70°	0.9397	250	234.9
80°	0.9848	250	246.2
90°	1.0000	250	250

Figure 10-5. The specific amount of voltage produced may be found by use of a sine table.

information, along with the amount of maximum voltage produced, is used to determine the voltage at each instant of time during one revolution of the armature. Use 20 degrees as a point of reference. The maximum voltage produced is 250 volts. The formula $E_{\text{inst}} = E_{\text{max}} \times \text{sine } \theta$ is used to provide the instanteous voltage of the generator. Using number values, $E_{\text{inst}} = 250 \times 0.34$, gives an answer of 85 volts produced at that instant of time. Generated voltage for every instant of time during rotation of the armature is calculated in this manner. Plotting the results of these calculations gives a sine curve.

AC TERMINOLOGY

Several terms are used to describe an ac voltage. These are illustrated in Figure 10-6 and described in the following paragraphs. These

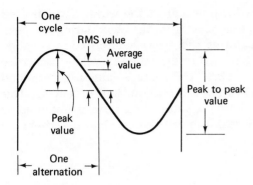

Figure 10-6. Specific terms relating to AC voltages are illustrated in this figure.

terms describe both ac voltage and ac current. They are used to provide a common reference for a constantly changing set of values.

Start with terms relating to time. One alternation of the armature turns it 180 degrees. This term is not as commonly used as is the term cycle. One cycle describes one complete revolution of the armature. The time reference in electrical work is the second. Armature rotation is described in units of rotation during a 1-second time period. The metric term describing cycles per second is hertz (Hz). This represents the frequency of repetition of the electrical wave during 1 second of time. The power line frequency in the United States is 60 hertz.

The period of the wave's duration is also a descriptive value. The period is the reciprocal of the frequency. Periods are described in units of the second. The time period for the ac power wave is $\frac{1}{60}$ of a second, or 0.0167 second. This may be used to convert from time to frequency by reversal of the formula. Frequency is the reciprocal of time. 0.0167 seconds divided into 1 gives an answer close to 60 hertz as the frequency of this wave.

The second set of terms describing the ac wave refers to the amplitude of the wave. This reference is for one cycle of the wave. The most common term used is one called the rms value. This is determined by squaring the values of voltage for each instant of time. These are then added together and divided by the number of instantaneous values used. The result is an average, or mean, value. The square root is then extracted from the mean value. This gives the rms, or root-mean-square, value of the ac wave. The rms value is also called the effective value of the ac wave. The rms value of the ac wave is normally used to describe it. In fact, when an ac value is given without any other descriptive terms, it is accepted as the rms value.

The other terms used to describe amplitude of the ac wave are peak, average, and peak-to-peak. The peak value, as illustrated, provides the maximum ac value for one half of the wave. E_{peak} is used to describe this value. The value of ac from the peak during the positive alternation to the peak during the negative alternation is called the peak-to-peak value. E_{p-p} is used to describe this value. It is twice that of the peak value of the wave. The average value is determined by averaging all instantaneous values during one half of a cycle. This value represents 0.636% of the peak value. This term is not used as often as the rms, peak, and peak-to-peak values when describing ac wave values.

CONVERSION FACTORS

The interelation of ac values may be determined mathematically. These are given in the table shown in Figure 10-7. Before attempting

Given	To find	Multiply by
E_{peak}	E_{rms}	0.707
E_{peak}	E_{avg}	0.636
E_{rms}	E_{peak}	1.414
E_{rms}	E_{avg}	0.899
E_{avg}	E_{peak}	1.57
E_{avg}	E_{rms}	1.11

Figure 10–7. The interrelation of the various terms describing the AC waveform may be found by use of these conversions.

to use these, let's see how they are derived. The rms value of the ac wave is 0.707 of the peak value. To convert from peak to rms, one multiplies the peak value by 0.707. Use the reciprocal of 0.707 to reverse this process. Multiply the rms value by 1.414 to find the peak value of voltage. The average values are figured in the same way. Here are some sample problems in order to help understand how to use these conversions.

1. Find the peak value of waveforms with an rms value of 230 volts.

$$E_{peak} = 1.414 \times E_{rms}$$
$$= 1.414 \times 230$$
$$= 325.22 \text{ volts}$$

2. Find the average value of a waveform with an rms value of 230 volts.

$$E_{av} = 0.636 \times E_{rms}$$
$$= 0.636 \times 230$$
$$= 146 \text{ volts}$$

3. Find the rms value of a waveform whose positive peak is 100 volts.

$$E_{rms} = 0.707 \times E_{pk}$$
$$= 0.707 \times 100$$
$$= 70.7 \text{ volts}$$

4. Find the average value of a waveform whose negative peak is 250 volts.

$$E_{avg} = 0.636 \times E_{pk}$$
$$= 0.636 \times -250$$
$$= -159 \text{ volts}$$

5. Find the peak value of a peak-to-peak waveform whose value is 780 volts.

$$E_{peak} = \frac{E_{p \cdot p}}{2}$$
$$= \frac{780}{2}$$
$$= 390 \text{ volts}$$

6. Find the frequency of a waveform with an rms value of 250 volts and a time period of 0.010 second.

$$f = \frac{1}{T}$$
$$= \frac{1}{0.010}$$
$$= 100 \text{ Hz}$$

Note: Amplitude of the waveform has no effect on its frequency.

7. Given a waveform whose frequency is 400 hertz, find the time period.

$$T = \frac{1}{f}$$

$$= \frac{1}{400}$$

= 0.0025 second

PHASE ANGLES

The term phase angle describes the relationship between the starting times of any two electrical waves. Each electrical power wave is considered to be produced from a generator. The generator rotates through a full 360 degree circle in order to create the sine wave form. The starting point of the wave is usually considered to be at a point midway between the two poles of the generator. This point is considered to be zero volts. Rotation of the armature creates the sine wave form. If a second armature were to be connected onto the same shaft as the first armature, two waves could be created as the armatures turn. If both armatures were in the same alignment, both sets of waves would start and end at the same moment in time. This is illustrated in Figure 10-8. Both of these waves start and end at the same

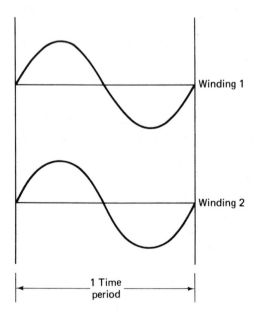

Figure 10-8. The phase relation of two AC waves is described by comparing the starting times of each wave.

time. They are said to be *in phase* with each other.

There is little reason to have two identical sets of armature windings on one shaft of a generator. In many cases any additional windings are rotated on the shaft so that they would reach the mid-point between the two poles of the field at a different moment in time. The resulting waveforms created by this situation are shown in Figure 10-9. Here the second waveform starts after the first one has com-

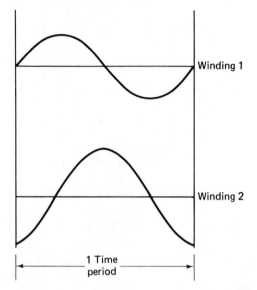

Winding 1

Winding 2

1 Time
period

Figure 10-9. Two out-of-phase waves. These waves are 90 degrees out of phase with each other.

pleted 90 degrees of rotation. The first wave is used as a reference. The second wave is said to be *out of phase* with the first. It lags behind the first wave. The description of the second wave is that it lags behind the first wave by 90 degrees. It is 90 degrees out of phase with the first, or reference, waveform. The amount of lag depends on the relationship of starting times during the period of rotation.

Any electrical wave may be described in this manner. Waves created by electronic means may also be described this way. The phase angle of the wave refers only to its relation to a reference wave. Waves may also lead. The second wave could start before the reference wave. It then would be leading the other wave by a number of degrees. It is also possible to describe the relation between voltage and current in a circuit by comparing the phase angles of each. Here, too, one could be leading the other by x degrees.

Many ac generators have three sets of armature windings. These sets are spaced 120 degrees apart on the shaft. This type of generator is called a three-phase generator. It is used to produce the electrical power requirements for home, industry, and automobile. Three-phase power is more efficient than single-phase power, because more power can be produced in 1 second with three sets of armature windings than can be produced with a single set of windings. These are illustrated in Figure 10-10.

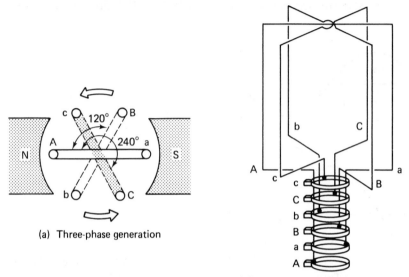

(a) Three-phase generation

(b) Output connections of generator

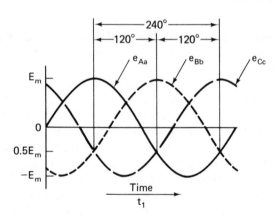

Figure 10–10.   Three windings on the armature of a generator produce a three-phase voltage.

Both single-phase and three-phase systems are used to deliver electrical power. The classification of the systems is simply the number of phases used. Therefore, the terms single phase and three phase are used when discussing power-delivery wiring systems.

Electric power uses frequencies of either 50 or 60 hertz. These are fixed frequencies. They are held constant by the use of a computer. These frequencies are used to control the speed of electric clock motors. It is one way to maintain accuracy of time. Television sets also use the power line frequency as a time reference.

There are times, however, when other power frequencies are used. The military, in some instances, uses a frequency of 400 hertz for electrical power. This higher frequency allows the utilization of smaller generators and transformers in order to produce the same amounts of power as 60-hertz systems provide. This is possible because more lines of flux are crossed during one rotation of the armature at higher frequencies. Thus, fewer wires are required to produce a voltage equivalent to that produced by 60-hertz systems.

This same principle is applied in many of the latest-technology television sets. One circuit in the set produces an electrical wave whose frequency is 15,734 hertz. This is the horizontal scanning frequency of a color television set. The waves from the scanning system are used to create operating power for other sections of the set. This system is called *scan-derived* because of the method of obtaining the power.

ELECTRICAL SIGNALS

The beginning of this chapter offered criteria for identifying power waveforms and signal waveforms. As our technology becomes more complex, the use of signal waves has increased greatly. Electrical signals are used to control electric power operation. These signals control the amount of power and the timing of the use of the power. In many cases the electric signal is added to wiring carrying electric power. Both are sent from the source on the same set of wires. There is no interaction between the two because of the difference in operating frequencies. The electrical technician needs to know about these electrical signal frequencies.

Electrical signals are divided into groups by those who use them. The two major groups are called *audio* frequencies and *radio* frequencies. The term used to describe audio frequencies is AF. Audio frequencies range from about 10 hertz to about 20,000 hertz. These frequencies may be heard by most human ears. Information carried by radio or tape recorders is produced in the audio-frequency range.

The group of frequencies between 20,000 and 30,000 hertz is called supersonic frequencies. This is because they are higher than audio (sonic) frequencies, but do not meet the criteria for the next group of frequencies. The pitch, or tone, of any sound is directly related to the frequency of the wave. As the tone becomes a higher pitch, the frequency of the waveform also increases.

Once the electrical signal becomes higher than 30,000 hertz it is called a radio frequency. This is because signals at this frequency, or higher may be effectively broadcast by a radio station's transmitting antenna. This group of frequencies ranges from 30,000 hertz to over 300,000,000,000 hertz. A large variety of services use these frequencies. They are assigned by international agreement for various uses. These include commercial radio and television broadcasting, public service, teletype, CB, and amateur radio.

Radio frequency, or RF, waves are subdivided into groups according to the length of the electrical wave. These include low frequency (LF), high frequency (HF), very high frequency (VHF), and ultrahigh frequencies (UHF). Some of these are familiar because the terms VHF and UHF are used to identify television broadcast station groupings. Channels 2 to 13 are in the VHF range and channels 14 to 83 are in the UHF range. The study of AF and RF signals is too complex to cover in this book. Actually, volumes have been and are being written covering each of the areas of audio and radio signals. The electrical technician should recognize that there are frequencies being used that differ from the power frequency of 60 hertz. These signals are carrying information of some type. The information may be used to transfer a message from the source to the receiver or to control a machine. It may be data input for a computer, or it may be the complex telemetry signal communicating between a space vehicle and mission control.

These signals tend to become much more complex in shape than does the power waveform. Some of these shapes are illustrated in Figure 10-11. Signals of these wave shapes are found in color television sets. Many other forms are used in communication and industry. A special piece of equipment, the *oscilloscope*, is required when attempting to observe and measure these waveforms. Its operation is discussed in Chapter 15. At this point the technician should recognize that there are power and signal sources developed from alternating current. These vary in shape, repetition rate, and amplitude. They are used for power operation or signal information purposes as required.

DIRECT CURRENT

The direct current (dc) generator is built in a manner similar to the ac generator. The basic difference between the two generators is

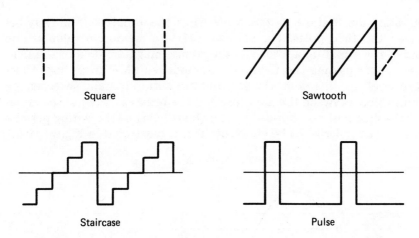

Square   Sawtooth

Staircase   Pulse

Figure 10–11. Electrical signals often have shapes that differ from the sine wave.

the method of removing the induced EMF from the armature. The dc generator uses a commutator instead of the slip rings found on the ac generator. The commutator is aligned on the armature shaft in such a way that the EMF at one brush is always positive. The EMF at the other brush of the generator is always negative. This is illustrated in Figure 10-12. The resulting output waveform from the generator is a

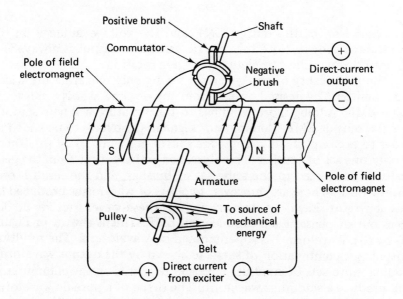

Figure 10–12. The use of a segmented commutator produces a DC voltage.

pulsating dc. It starts at zero volts when the armature is midway between the field poles. The voltage rises to a maximum value as one end of the armature approaches the intense magnetic field of the field pole. After passing the field pole, the induced EMF drops back to its zero level. At this point the commutator action literally switches the connection between the armature and the brushes. Now, as the armature swings past the opposite pole, the reversal of the wiring permits the output polarity to be constant. This is illustrated in Figure 10-13.

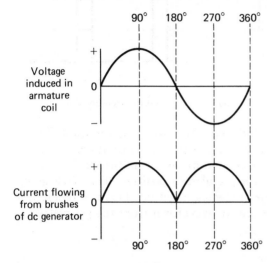

Figure 10-13. A comparison of an AC wave and a DC wave of a generator.

A comparison of the output EMF and the voltage induced on the armature is shown. Note that the dc generator output is always one polarity through the complete 360-degree rotation.

The practicality of any generator having only one armature winding is limited. The desired output from any dc generator is a constant-value EMF. Adding more windings to the armature will help smooth out the output EMF. Commutator segments also have to be added in order to accomplish this. These are shown in Figure 10-14 (a). There is only one set of brushes. Therefore, the entire output from the generator is connected to the same set of brushes, and the result is one output EMF. There are, however, two sets of waveforms produced in this generator. Each has its own timing. These waveforms are 90 degrees out of phase. Each produces an EMF. This is shown in Figure 10-14 (b). Waveform 1 is superimposed on waveform 2. The resulting output is a combination of these, as shown by the output waveform. Adding more sets of armature coils and segments on the commutator will produce a smoother waveform. The shape of a pure dc waveform is a straight line. This shape may be accomplished by additional armature segments.

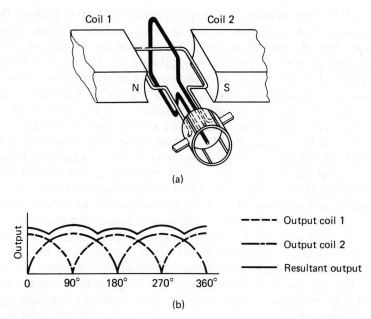

Figure 10–14. A two-phase generator produces a wave that is 90 degrees out of phase with its other wave.

The percentage of dc generators when compared to ac generators is very small. Most installations of generators use the ac generator, or alternator, because of its efficiency. In automotive applications the alternator has replaced the generator. The size of the alternator is much less, and its comparative weight is less. Added to this is the greater efficiency of the alternator. The automobile requires dc to operate its various circuits. The alternator output is changed into dc for use in the car. The changing of ac into dc is called *rectification*. This process and related material are presented in Chapter 18.

SUMMARY

Operational power sources are almost exclusively ac, which is used because it is easier to produce than dc. Alternating current has a varying waveform. It may vary in amplitude, frequency, and shape. There are two types of ac waves. These are generally called signal and power waves. Each is identified by relation to its application and frequency.

Applications of ac waves for power usually require large currents and voltages. Smaller requirements are classified in signal waves. Signal waves are information-carrying waves. Their frequencies are normally very high. Power frequencies are normally around 60 hertz.

Another classification for power is called dc. Direct current flows in only one direction in the circuit. In many instances the ac power is changed into dc power in order to operate an electronic device.

Electric power is found by use of the power formula $P = E \times I$. It can be seen that increasing either the voltage or the current values will increase the amount of power. If one wishes to maintain a fixed power level, then a rise in voltage is offset by a drop in current. This factor is used in the transportation of electric power. Voltage values are increased to a great extent. Current values will drop if the power level is held constant. High-voltage and low-current power requires smaller sized conductors than those required if the current was a higher value. There is much less of a voltage drop and resulting power loss when a low level of current is transported at a high voltage.

A sine wave is produced as an armature revolves in a magnetic field. The voltage developed by electromagnetic induction is dependent upon the relationship of the armature to the magnetic poles in the field. The greatest strength of induced voltage occurs when the armature is passing the pole. The amount of voltage induced on the armature at any moment in time is determined from the formula $E_{inst} = E_{applied} \times \text{sine } \theta$.

Terminology is important when discussing ac values. An alternation is a 180 degree rotation of the generator armature. A cycle is one complete revolution of the armature. The time reference period for electrical concepts is 1 second. Frequency of the wave refers to the number of complete cycles in 1 second. The term hertz is used to describe this occurrence.

Another way of describing the ac wave is to measure its amplitude. The most common way is called the rms value. Another way is to measure its peak-to-peak value. Sometimes the peak value is used. The peak-to-peak value is 2.828 times the rms value of the wave.

The phase angle of the ac wave refers to the starting time of one wave compared to the starting time of another wave. Each ac wave has a starting point in a rotational field of 360 degrees. This is directly related to the rotation of the armature of the generator. Zero degrees is usually located at the mid-point between the generator's poles. Two waves with different starting times are described as having a phase shift of x degrees. The exact number of degrees depends upon how much after the first wave the second wave starts. This reference is also true with electronic wave generators.

Electric signals usually have a frequency that is much greater than the 50- or 60-hertz power frequency. Audio frequencies range up to 20 kilohertz. Radio frequencies are those frequencies above 30 kilohertz. These signals are observed by means of an oscilloscope. It presents a visual image of voltage waves on the screen of a cathode-ray tube.

Direct-current power is also produced by a generator. The difference between ac and dc generators is basically the method of removing the power from the device. A dc generator uses a segmented commutator to maintain circuit polarity. Several armature windings are connected to a pair of segments in the commutator. This increases the efficiency of the generator and produces a wave that closely resembles pure dc form.

FORMULAS

Frequency: $f = \dfrac{1}{T}$

Instantaneous ac voltage: $E_{inst} = E_{max} \times sine\ \theta$

Rms value: $E_{rms} = E_{peak} \times 0.707$

Average value: $E_{av} = E_{peak} \times 0.636$

Wave time: $T = \dfrac{1}{f}$

Peak-to-peak value: $E_{p\text{-}p} = E_{peak} \times 2$

PROBLEMS

10.1. Find the time period for the following frequencies:
   a. 60 Hz
   b. 1,000 Hz
   c. 1,500,000 Hz
   d. 12,000 Hz

10.2. Find the frequency for each wave time:
   a. 10 s
   b. 0.040 s
   c. 0.000060 s
   d. 0.16 s

10.3. Find the average value of an ac wave with a 120-V peak rating

10.4. Find the peak-to-peak value of a 120-V rms wave.

10-5. Find the peak value of a 480-V rms wave.

10-6. Find the rms value of a 440-V peak wave.

10-7. Find the rms value of 1,000-V peak-to-peak value.

10-8. Find the peak value of a 220-V rms wave whose frequency is 60 Hz.

10-9. Find the average value of a 440-V peak wave.

10-10. Find the average value of a 220-V peak-to-peak wave.

QUESTIONS

10-1. Voltages used in the electric power industry are almost always

    a. dc

    b. RF

    c. ac

    d. signal

10-2. The power line frequency used is

    a. 25 to 30 Hz

    b. 5 to 60 Hz

    c. 60 to 100 Hz

    d. 100 to 120 Hz

10-3. The shape of the electrical wave is called a

    a. square wave

    b. triangular wave

    c. signal wave

    d. sine wave

10-4. One complete revolution of the generator armature is called

    a. a hertz

    b. an alternation

    c. a frequency

    d. a cycle

10-5. RMS values are ____ of peak values

    a. 1.414

    b. 2.828

    c. 0.707

    d. 0.634

10-6. The difference in rotational degrees of starting of two electrical waves is called the

    a. phase angle

    b. rotation delay

    c. starting difference

    d. none of the above

10-7. Electrical signal frequencies are

    a. 25 to 50 Hz

    b. 100 to 20,000 Hz

    c. 60 Hz

    d. above 20 kHz

10-8. Match the term to the wave:

    A. ____   1. rms

                            2. peak

    B. ____   3. effective

                            4. peak to peak

    C. ____   5. cycle

                            6. frequency

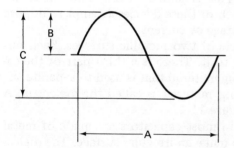

11

Capacitance

The three basic factors that influence any electrical circuit were identified earlier. They are resistance, inductance, and capacitance. Both resistance and inductance are discussed in earlier chapters. The final factor, capacitance, is discussed in this chapter. The capacitor reacts to changes in circuit *voltage*. This is similar to the action of an inductor with circuit *current*. Both of these devices attempt to oppose any change in the respective voltage or current.

The simple capacitor consists of two metallic surfaces. Each surface has a wire lead connected to it. There is a third part of the capacitor. This part is an insulating material that is used to separate the metallic surfaces. This insulating material is called the *dielectric*. A simple capacitor is shown in Figure 11-1.

The metalic surfaces used in most capacitors are made of metal foil. The foil is often etched to make an irregular surface. The dielectric material is any nonconducting material. Some capacitors use air as the dielectric. Other materials in common use as dielectrics include paper, ceramic, mica, oil, and plastic. The purpose of the dielectric is to electrically separate the two metal surfaces of the plates.

Capacitors have been, and are still, called *condensers* in some electrical work. The name condenser came about because scientists felt that the capacitor had the ability to condense electricity. This theory is not true. The fact of the matter is that the capacitor has the capacity to store an electrical charge. From this revised and correct theory comes the name capacitor.

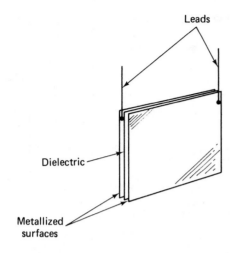

Leads

Dielectric

Metallized
surfaces

Figure 11-1. The basic capacitor is made of two metal surfaces
separated by a dielectric material.

CAPACITOR ACTION

The action of the capacitor is illustrated in Figure 11-2. The capacitor is shown in three of the four drawings in this illustration. The
other part shows the schematic symbols for both fixed (nonadjusta-
ble) and variable capacitors. Start with the assumption that the capacitor is not charged. Each of the two plates is electrically neutral.
They contain a balanced number of positive and negative charges.
Part (a) shows the charging action of the capacitor. A voltage source
is connected to the plates of the uncharged capacitor. When this is
done, there is a movement of electrons in the circuit. The bottom
plate collects electrons. Electron flow is from the source to the plate
of the capacitor. This gives the capacitor plate a negative charge. At
the same time that this is occurring, the electrons on the upper plate
are attracted to the positive terminal of the source. When the electrons leave the upper plate, it becomes positively charged. The factors
that determine the amount of charge are discussed in the next section. The flow of electron current continues until each of the two
plates is fully charged.

The capacitor will remain charged until some condition makes it
give up. The source that created the charge may be removed and the
capacitor plates will remain in a charged condition. This is illustrated
in part (b). If it is necessary to discharge the capacitor, it may be accomplished as shown in part (c). A switch is connected across the capacitor terminals in this model. In a practical circuit the switch is not
used. It is replaced by a portion of a functional electric circuit. When

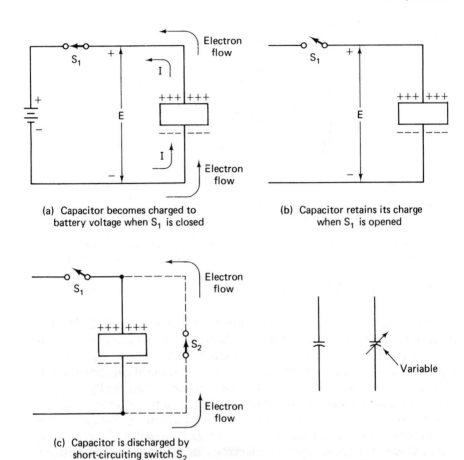

(a) Capacitor becomes charged to
    battery voltage when S_1 is closed

(b) Capacitor retains its charge
    when S_1 is opened

(c) Capacitor is discharged by
    short-circuiting switch S_2

Figure 11-2.   Capacitor action is illustrated. The plates are charged by the source
(a). The charge is held (b) when the source is removed. Capacitor is discharged when
it is connected to a circuit (c). Schematic symbols for capacitors (d).

either the switch or the circuit is turned on, the current will flow
through the switch. Current flow is from the negative charged plate
to the positive charged plate. The electrons travel to the positive plate
and neutralize its field. As the positive plate is neutralized, there is a
movement of positive charges through the circuit to the negative
plate. This tends to neutralize the original charge on this plate. The
capacitor is now discharged.

Capacitor action is very similar to the action of a spring. The
spring at rest has no energy. If the spring is compressed, it stores energy. The spring tends to oppose being compressed. It has a natural
tendency to return to its "at rest" condition. If the spring is stretched,
it will also attempt to oppose the stretching action. It tries to return

to its "at rest" condition from the stretched position. The spring acts against mechanical forces.

The capacitor acts against electrical forces in the same manner. When a capacitor is connected into an electrical circuit, the capacitor tends to oppose any changes in voltage. As the voltage drops, the capacitor gives up its charge in order to try to maintain its "at rest" condition. For the capacitor, this position means that it has a charge equal to the voltage in the circuit before any changes occur. On the other hand, if the circuit voltage tries to rise, the capacitor will attempt to return the voltage to the "at rest" condition. This action is limited by the capacity of the capacitor to hold a specific amount of charge.

CAPACITANCE FACTORS

The factors that affect the amount of charge on a capacitor are related to size, spacing, and the dielectric materials. The size of the metallic plates is directly proportional to the amount of the charge on the plates. Increasing the surface area of the plates increases the capability of charging each plate. The capacity of the electrical charge is also increased. The second factor involved in the amount of capacitance relates to the spacing between the two metallic surfaces. Charges tend to act upon each other in any circuit or device. The capacitor plate that is connected to the negative terminal of the source will become negatively charged. This negative field will interact with the neutral field of the other plate. It tends to drive electrons away from the second plate. Interaction between any two magnetic fields is related to the distance between these two fields. This is also true in the capacitor. If the fields established in the metallic plates are far apart, there will be little interaction. The amount of capacitance in a given capacitor, therefore, is dependent upon the spacing between its two plates.

The final factor in determining the amount of capacitance is related to the dielectric material used in the capacitor. The strength of any dielectric insulating material is related to the material's ability to withstand a voltage breakdown. During voltage breakdown an arcing occurs and electrons literally jump from one surface to another. Dielectric strength is measured in the number of volts that are protected for each mil of thickness. Air has a dielectric factor of 1. Its dielectric strength is about 75. An EMF of 75 volts is required to create an arc between two surfaces that are spaced in air 1 mil apart. Other materials have higher dielectric factors. Glass has a dielectric strength of about 3,000. Mica's dielectric strength is over 5,000. The selection of the proper dielectric material enables the manufacturer of capacitors to make very small units that are able to withstand fairly high voltages.

One set of markings on the capacitor's case relates to its *working voltage*. This term refers to the value of voltage the capacitor's dielectric will insulate against. If a capacitor with a working voltage of 450 is placed in a circuit whose voltage is higher than 450, the voltage will probably arc between the plates of the capacitor. This will ruin the dielectric and often develop into a short-circuit condition within the capacitor.

The operating voltage of the circuit is a major factor to consider when selecting a capacitor. The working voltage of the capacitor must be higher in value than any voltages found in the circuit in which the capacitor is to be used. It is possible to install a capacitor whose voltage rating is two or more times the operating voltage as long as there is physical space available for the capacitor. A general rule applies to physical size and working voltage for a specific type of capacitor. These two values are usually directly proportional to each other. If one had two capacitors that had the same capacitive value and were formed with the same type of plate and dielectric material, then the capacitor with the higher voltage rating would be larger physically than the capacitor with the lower voltage rating. Keep in mind that if two capacitors are constructed using different dielectric and plate materials, no size relationship rule applies.

UNITS OF CAPACITANCE

The unit of capacitance is the farad (F). This is an extremely large unit and is not used very often. Most capacitors are rated in small fractions of the farad. The terms normally used to describe the amount of capacitance are the microfarad and picofarad. One microfarad (μF) is equal to one millionth of a farad, or 0.000,001 farad. One picofarad (pF) is equal to one millionth of one millionth of a farad, or 0.000,000,000,001 farad. These two terms are normally used in the field to describe quantities of capacitance. Capacitors are identified on the circuit or parts list by the letter C.

TYPES OF CAPACITORS

The sizes and shapes of capacitors are very numerous. Certain characteristics for capacitors enable one to categorize them. Capacitors often fall into groups. The groups relate to the construction of the capacitor. A discussion of each general type will help with the terminology. Typical capacitors are illustrated in Figure 11-3.

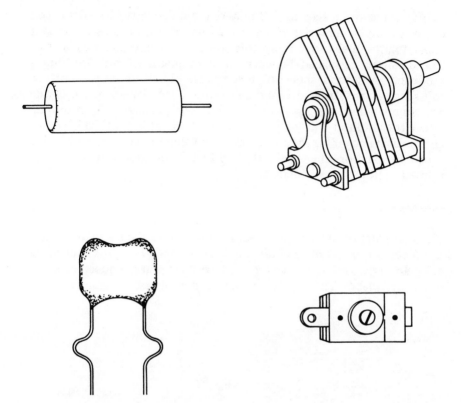

Figure 11–3. Capacitors have a variety of shapes. Two fixed-value units are on the left and two variable value units are shown on the right.

Tubular Capacitors

One of the most common types of capacitor is the tubular capacitor. Its construction is shown in Figure 11-4. This capacitor is made of

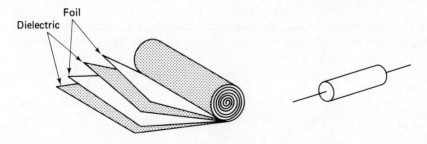

Figure 11–4. Tubular capacitor construction.

two long narrow pieces of foil material and dielectric. The dielectric often is a mylar material. Early capacitors of this type used a waxed paper for the dielectric. From this comes the name paper capacitor. The dielectric is placed between the two pieces of foil, forming a sandwich. A second sheet of dielectric is placed under the bottom foil strip. These strips are then rolled into a tubular form. Wire leads are connected to the two foil surfaces. The package is then dipped into either molten wax or plastic. When hard, this material becomes an insulating cover for the capacitor. The value of capacitance for this type of capacitor ranges from 0.0005 to 1.0 microfarad. This value is printed on the case of the unit.

Mica Capacitors

The construction of a mica capacitor is shown in Figure 11-5. Mica is the dielectric material used for these capacitors. Mica-type capacitors are made in layers as illustrated. The wire leads are connected to the respective plates. The unit is then encased in plastic. One variation of

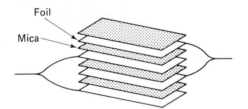

Foil

Mica

Figure 11-5.   Mica   capacitor construction.

the mica capacitor is called silvered mica. The mica dielectric has silver deposited on it instead of using metallized foil for the plates. Mica capacitors are closer in tolerance to their marked value than those made from other materials. Capacitance range for mica capacitors is from 1 picofarad to 0.1 microfarad.

Ceramic Capacitors

Ceramic capacitors are made using ceramic as a dielectric. Metal is deposited on each side of a disc of ceramic. Wire leads are connected

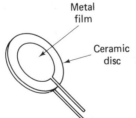

Metal film

Ceramic disc

Figure 11-6.   Disc capacitor construction.

to each of the metal surfaces. This is illustrated in Figure 11-6. The assembly is then covered with a plastic coating. This forms the ceramic capacitor. Values for ceramic capacitors range from 1 picofarad to 1.0 microfarad.

Electrolytic Capacitors

When values greater than 1.0 microfarad are required, an electrolytic type of capacitor is used. The basic construction of the electrolytic capacitor is similar to that of the tubular capacitor. Two sheets of aluminum foil are separated by a cloth dielectric material. The dielectric is first soaked in an electrolyte. The three layers are then rolled and placed in a tubular metal container. After assembly, a dc voltage is connected to the two plates, causing a thin layer of aluminum oxide to form on the positive plate. The aluminum oxide becomes the dielectric. The electrolyte and the positive foil sheet are the plates of the capacitor. This produces some very large values of capacitance.

This is an excellent place to insert a remark about safety when using electrolytic types of capacitors. These capacitors have polarity markings on either the metal container or the leads. Correct polarity must be observed when connecting an electrolytic capacitor into a circuit. If the polarity is not observed and reverse polarity connections are made, an explosion may occur. The chemicals in the electrolytic capacitor will form a gas. The gas expands in the container. It has no place to go, so it will explode the container. Flying metal is hazardous.

Electrolytic capacitors are available in ranges from 1.0 microfarad to well over 200,000 microfarads. They are rated both in capacitance value and in working volts. The tolerance of the electrolytic capacitor is fairly broad. It is often possible to use a replacement electrolytic capacitor whose value is as much as 50% above the original value. Working voltage values must be equal to or greater than the original value when replacing any capacitor.

Variable Capacitors

All the capacitors described so far are classified as *fixed* capacitors. Their value has no way of being changed. There is a second group of capacitors. These are called variable capacitors. The amount of capacitance of a variable capacitor may be adjusted. Two basic types of variable capacitors are shown in Figure 11-7. The ceramic trimmer capacitor consists of two ceramic plates. Each plate has a metallized surface on part of the plate. The upper plate rotates by means of turning it with a screwdriver. When the plates are on top of each other, the capacitor has its greatest value of capacitance. Rotating the upper

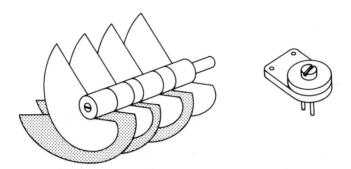

Figure 11-7.   Variable capacitor construction.

plate 180 degrees provides a minimum amount of capacitance.

The second type of variable capacitor is made of two groups of plates. One group of plates is mounted in a frame. The plates are usually insulated from the frame. The second group of plates is attached to a rotating shaft. When the two sets of plates are meshed, the interaction between them is the greatest. This provides a relatively large amount of capacitance. When the plates are not meshed, the amount of capacitance is at a minimum. The capacitor is adjusted by turning the shaft and changing the relationship of the two sets of plates.

CAPACITORS IN SERIES AND PARALLEL

Capacitors may be connected in either series or parallel. This is done when a value of capacitance other than that available is required. A review of how a value of capacitance is established will aid in understanding what happens to total capacitance when two or more capacitors are wired together.

Earlier in this chapter the three factors for determining capacitive values were identified. They related to area of the plates, spacing between plates, and dielectric material. Increasing the total area of each plate of a capacitor will increase the amount of capacitance. When capacitors are wired in parallel, the total plate area connected to the circuit is increased. This is shown in Figure 11-8. Each capacitor's plate area adds to make a total plate area. The capacitance as measured across any of the capacitors will be equal to the sum of the values of each capacitor. This is expressed in the formula

$$C_T = C_1 + C_2 + C_3 \text{, etc.}$$

Capacitors wired in series offer a different set of values to the circuit. The distance between those plates that are connected to the circuit determines the total effective capacitance. Spacing between

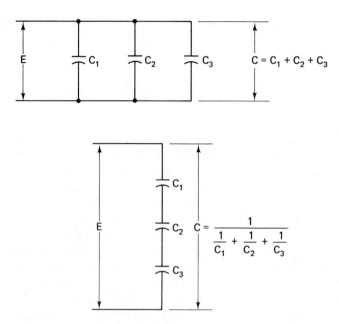

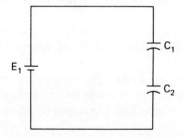

Figure 11-8. Wiring combinations for capacitors. The formulas for finding total capacitance in each circuit are shown.

plates of a capacitor helps to determine the value of capacitance. Figure 11-9 illustrated this point. Two capacitors are wired in series. The

E_1

C_1

C_2

Figure 11-9. A series-wired capacitor circuit.

upper plate of C_1 is connected to the positive terminal of the power source. The lower plate of C_2 is wired to the negative terminal of the power supply. The other plates of each capacitor are wired together, but do not connect to a power source. The charges on these plates are developed from those plates connected to the power source. Charges on the upper and lower plates determine the total capacitance. The effecitve capacitance for series-wired capacitors decreases in this circuit arrangement. The total capacitance is found by adding the reciprocals of each capacitive value. The formula for series-wired capacitors is

$$C_T = \cfrac{1}{\cfrac{1}{C_1} + \cfrac{1}{C_2} + \cfrac{1}{C_3}}, \text{etc.}$$

SUMMARY

Capacitance is one of the three basic factors that affect electronic circuits. The capacitor is constructed of two metallic plates, and an insulating material, called the dielectric, is sandwiched between the two metal plates of the capacitor. Dielectric materials used in capacitors include air, mica, oil, paper, plastic, and ceramic.

The action of the capacitor is like that of a spring. The capacitor tends to oppose any change in the amount of voltage charge on its plates. The capacitor will give up charges as the circuit voltage drops. It will also try to absorb charges as the circuit voltage rises.

Capacitor values are usually given in microfarads or picofarads. The size of one unit of capacitance, the farad, is too large for formal electrical circuit work. Fractions of the farad, therefore, are used. The capacitor is rated in subunits of the farad. It is also given a working voltage rating. The working voltage rating shows a maximum usable circuit voltage. Voltages higher than the rated voltages will usually damage the capacitor.

The factors involved in determing the amount of charge for a capacitor depend upon the area of the plates, the spacing between the plates, and the type of dielectric material used. These three factors are interrelated.

The most common types of capacitors include tubular, ceramic, mica, and electrolytic. Each name describes the type of dielectric material used in the capacitor.

Most capacitors are not polarity sensitive. The exception to this is the electrolytic capacitor. If correct polarity is not observed, the chemicals in the electrolyte form a gas. The gas expands inside the metal container of the capacitor. The expanding gas has no place to go, so an explosion occurs. Always observe correct polarity when connecting an electrolytic capacitor.

Not all capacitors have fixed values. Some capacitors are made so that the amount of capacitance may be varied. This group is called either variable or trimmer capacitors. In either case, the amount of capacitance is varied by moving one plate with respect to a fixed plate. As the plates move away from each other, the amount of capacitance decreases.

Capacitors may be wired in parallel. When they are wired in this manner, the total amount of capacitance increases. It is then equal to the sum of each value of capacitance.

Capacitors may also be wired in series. The resulting capacitance for this circuit arrangement is a much smaller value, because the plates that accept the charges are farther apart. Series-wired capacitance is equal to the sum of the reciprocals of each capacitor's value.

PROBLEMS

Find the total capacitance for the following capacitive circuits:

11-1.   1 μF, 10 μF, 6.8 μF, parallel wired.

11-2.   5 μF, 0.06 μF, 1,000 pF, parallel wired.

11-3.   0.04 μF, 120 pF, 68 pF, 15,000 pF, parallel wired.

11-4.   4 μF, 10 μF, 8 μF, series wired.

11-5.   120 pF, 60 pF, 180 pF, series wired.

11-6.   100 μF, 2 μF, 0.05 μF, series wired.

QUESTIONS

11-1.   The unit of capacitance is the

    a. coulomb

    b. volt

    c. ohm

    d. farad

11-2.   The insulating material used in the construction of a capacitor is called the

    a. insulating layer

    b. dielectric

    c. nonconductive layer

    d. charge separator

11-3.   One micro unit is equal to

    a. 0.001 unit

    b. 0.0001 unit

    c. 0.00001 unit

    d. 0.000001 unit

11-4.   Capacitor value depends upon

    a. plate size

    b. insulation material

   c. plate spacing

   d. all of the above

11-5. Insulating materials have

   a. equal properties

   b. different properties

   c. nonlinear properties

   d. none of the above

11-6. Capacitors in series add capacitance because of

   a. increased plate size

   b. increased area

   c. increased plate spacing

   d. none of the above

11-7. Capacitors in parallel add capacitance because of

   a. decreased plate size

   b. decreased plate area

   c. increased plate area

   d. none of the above

11-8. A variable capacitor, when fully meshed, has

   a. maximum capacitance

   b. minimum capacitance

   c. variable capacitance

   d. none of the above

11-9. Electrolytic capacitors

   a. are polarity sensitive

   b. are not affected by applied voltage polarity

   c. will reverse charge with reverse polarity voltage

   d. are not polarity marked on their housings

11.10. Electrolytic capacitor values range from

   a. 0.000001 to 0.001 units

   b. 0.0001 to 0.1 units

   c. 0.01 to 10.0 units

   d. 1.0 to 10,000 units

12

Reactance and Impedance

Each electrical component described in this book will exhibit some properties of resistance, inductance, and capacitance. Very often the amounts of capacitance and inductance in a resistor are minimal. These amounts do not have much of an effect on the electrical circuit. There are other times when either capacitance or inductance are deliberately included in the circuit. It is these conditions that are of concern to the electrical technician. The effects of capacitance, inductance, and resistance in the circuit are discussed in this chapter.

PHASE RELATIONSHIP

One of the terms commonly used in describing electrical circuits is the phase relationship of voltages and currents. Electrical phase is directly related to the timing of the electrical waves. When two voltages start at the same moment in time, they are *in phase* with each other. If the two waves do not start at the same moment in time, the waves are *out of phase* with each other.

The creation of the sine wave is the result of the rotation of a wire through 360 degrees of a magnetic field. This rotation of the armature wire provides a direction relationship between the position of the wire in the armature and the degrees of a circle. Figure 12-1 illustrates this point. As the armature wire rotates through one complete cycle, a voltage is created on the wire by the action of induction. The voltage varies depending on the position of the wire. Each moment in time is related to a degree of rotation through one complete cycle of

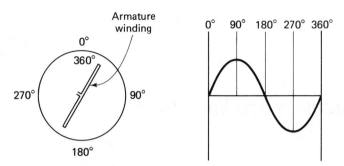

Figure 12-1. Development of the sine wave as the armature winding rotates through one complete 360-degree cycle.

the armature. From this the phase relationship of electrical waves is developed.

Two separate armature wires are placed on the rotating shaft of the generator. There will be two output voltages created. One output is labeled *A* in Figure 12-2. The second output is labeled *B*. The two

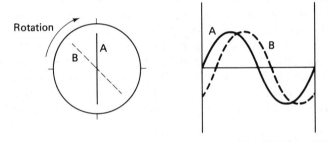

Figure 12-2. Two armature windings that are out of phase will produce two out-of-phase voltages.

wires are separated by 45 degrees of rotation. Wire *A* passes the 0 degree point before wire *B*. The difference in rotational time for wire *B* to cross the same point is 45 degrees. Wire *B* is said to have a phase shift of 45 degrees. It lags behind wire *A* by this rotational amount.

The relationship of the two voltages is always described by the degree of separation of the two voltage waveforms. One is either leading or lagging with respect to the other waveform. This also is true when describing the relationship between two currents. It is also true when discussing the relationship between voltage and current in a circuit.

Resistive Circuits

The circuit shown in Figure 12-3 is a purely resistive circuit. Any amounts of either capacitance or inductance are purely negligible. In

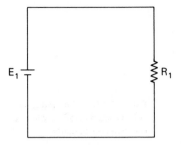

Figure 12-3. A resistive circuit contains only resistive components.

this type of circuit, voltage and current are in phase. The circuit lacks any of the properties that would tend to cause a shift in phase between voltage and current.

Capacitive Circuits

Should the circuit be modified by adding a capacitor, as shown in Figure 12-4, then the action of the capacitor will affect voltage and cur-

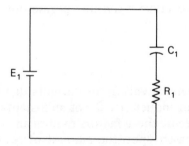

Figure 12-4. A capacitive circuit contains both capacitive and resistive components.

rent. Capacitors tend to oppose any change in circuit voltage. These changes take time, as does the opposition to the change. This being true, voltage tends to lag behind the current in a circuit containing capacitance. Another way of stating this is that the current leads the voltage in a capacitive circuit.

DC Inductive Circuits

A different set of conditions exists in a circuit containing both resistance and inductance. In this type of circuit the inductor affects the current flow rather than the voltage. Inductive circuits, like the circuit shown in Figure 12-5, tend to oppose changes in the circuit current. If the circuit was purely resistive, Ohm's law would apply. Changes in voltage would apply. Changes in voltage would cause an immediate change in circuit current. Inductors cause a lag in changes in circuit current. Because of this lag, inductive circuits have the current lagging behind the voltage, or voltage will lead circuit current.

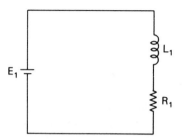

Figure 12-5. An inductive circuit contains both inductive and resistive components.

There is a way of remembering the phase relationship of voltage and current in a circuit. The way is to use the phrase, "ELI the ICE man." Break this phrase into electrical terminology. *E* represents voltage, *I* represents current, *L* is used to show inductance, and *C* represents values of capacitance. When the order of the terms is considered, the letters ELI mean that voltage leads the current in an inductive circuit (voltage-inductance-current). The second half of the phrase refers to the capacitive circuit. ICE shows that the current is leading the voltage when the circuit is capacitive (current-capacitance-voltage). The other words are only used to make a sensible phrase out of the terminology.

TIME CONSTANTS

The reaction to changes in circuit voltage or currents due to the influence of either capacitance or inductance is not an instantaneous occurrence. In either type of circuit these factors require an amount of time. The time required is based upon the amount of resistance and the amount of either capacitance or inductance. This is called the *LC* or the *LR* time-constant factor. The time constant is based upon the charging rate of the components in the circuit.

A circuit containing only resistance will react to an application of voltage in an almost instantaneous manner. The current curve for this type of circuit is shown in Figure 12-6. As soon as the current starts

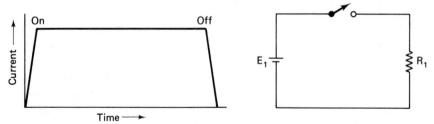

Figure 12-6. Current flow in a resistive circuit rises to the Ohm's law value. It remains at that level until power is removed.

to flow, it rises to its Ohm's law value and remains at that level. When the switch controlling current flow is turned off, the current immediately drops to zero.

This is not true in the inductive circuit. The action of the inductor causes a delay in the use of the current in the circuit. Current will rise to 63.2% of the Ohm's law value for one time constant. Time constant is determined from the formula $t = L/R$. t is equal to time in seconds, L is in henries, and R is in ohms. The charging rate for almost 100% is equal to five time constants. This is illustrated in Figure 12-7. The explanation for the changing curve is that the current

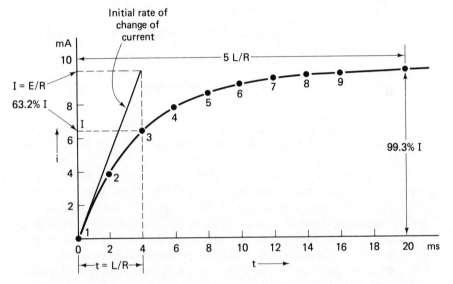

Figure 12-7. Charge time in an inductive circuit. The charge rises to 63.2% during each time-constant period.

rises 63.2% of the remaining level during each time constant. Thus, at the end of five time constants the current has reached 99.3%. This is close to 100%. An additional time constant unit's change is insignificant and not of sufficient value to be included. This type of curve is called an exponential curve. The formual for determining charging time in an LR circuit is:

$$t = 5\frac{L}{R}$$

The same rules apply to the discharge rate in an LR circuit. The graph in Figure 12-8 shows the curve developed when the current flow in an inductive circuit is turned off. The current falls to zero

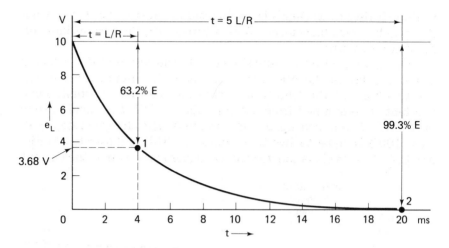

Figure 12-8.   Discharge time in an inductive circuit follows the same pattern as does charge time.

through five time-constant periods. This curve is the same shape but of opposite value to that used to illustrate the charging of the inductive circuit.

Voltage in a capacitive circuit will react in almost the same manner as does current in the inductive circuit. The capacitive component in the circuit opposes any changes in the amplitude of the circuit voltage. A certain amount of time is required to either charge or discharge a capacitor. This changing action works in the same manner as in an inductive circuit. Five time constants are required in order to have a full charge or discharge. The formula for determining the time constant of a capacitive circuit is

$$t = RC$$

where t is the time constant in seconds, R is the resistance in ohms, and C is the capacitance in farads. To find the total charge time, this formula is used:

$$t = 5RC$$

Curves showing both charge and discharge rates are shown in Figure 12-9. The only major difference between the two sets of curves is that the set for inductive circuits shows current on the vertical leg, while the set for capacitive circuits shows voltage on the same leg.

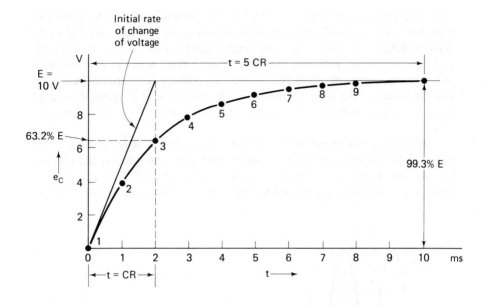

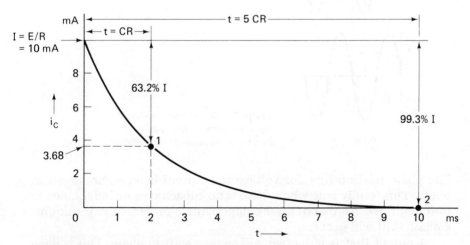

Figure 12-9. Capacitive reaction to changes in voltage follow the same patterns as do those for current in the inductive circuit.

AC RESISTIVE CIRCUITS

Voltage and current in any purely resistive circuit are in phase with each other. The purely resistive circuit action is the same whether a dc voltage or an ac voltage is applied to the load resistance. The current will rise to an Ohm's law value because of the action of the applied voltage. The dc voltage will cause the current flow. The current will rise in phase with the rise of voltage. Both voltage and current will remain constant until the source of voltage is removed from the load resistance. At that moment, both voltage and current will drop to zero. They will remain at zero until the voltage source is reconnected.

AC voltage is a constantly changing value. The ac wave is sinusoidal in form. When an ac voltage is applied to a resistive circuit, the changing voltage forces a changing current. Both rise, fall, and reverse direction at the same moments in time. This is shown in Figure 12-10.

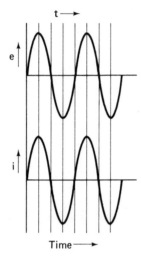

Figure 12-10. Current and voltage waveforms in the resistive circuit are in phase with each other.

The phase relationships for voltage and current in an ac circuit are in phase. This is only true when there is no capacitance or inductance in the circuit. When either of these components is present in any amount, a phase shift will occur.

A circuit that is inductive will cause a shift in phase. This is illustrated in Figure 12-11. The waveforms for the source voltage and the current through a theoretically purely inductive circuit show a 90-degree shift of phase. The voltage starts at time 0 and continues to time 4. This represents one cycle. The current starts at a later time. This time represents 90 degrees of rotation of the generator; thus the 90-degree phase shift between the two values.

Inductive Reactance

The effect of the inductance on the circuit is called the inductive re-

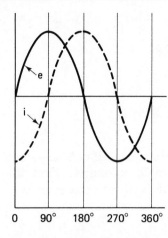

0   90°   180°   270°   360°

Figure 12-11. Current lags behind the voltage in a capacative circuit.

actance. This value is used to describe the opposition to current flow in a circuit due to the inductance in the circuit. The letter used to represent any type of reactance is the capital letter X. A subscript is used with the X to show that the reactance is inductive in nature. This gives the symbol X_L, which is measured to represent values of inductive reactance. X_L is measured in ohms.

A formula is used to determine the exact amount of inductive reactance. The ac wave, being a constantly changing value, has additional characteristics. These include frequency and the degree of rotation in the circular generator path. The formula for determining inductive reactance is

$$X_L = 2\pi f L$$

X_L is figured in ohms, f is the frequency in hertz, and L is the inductance in henrys. This formula is used to determine the exact amount of inductive reactance. The inductive reactance value, being in ohms, may be substituted for an ohmic value in solving for circuit current in the ac circuit. The two following examples show how this is accomplished.

Example 1. A 1.5-henry inductor is connected to a 120-volt, 60-hertz source. Find the inductive reactance and the ac current.

$$X_L = 2\pi f L$$

$$= 2 \times 3.14 \times 60 \times 1.5$$

$$= 5{,}652 \ \Omega$$

$$I = \frac{E}{X_L}$$

$$= \frac{120}{5652}$$

$$= 0.021 \ \text{A} \quad \text{or} \quad 21 \ \text{mA}$$

Example 2. An inductor is connected to a 12-volt ac source whose frequency is 15.75 kilohertz. Find the value of inductance when the circuit current is 100 milliamperes.

$$X_L = \frac{E}{I}$$

$$= \frac{12}{0.1}$$

$$= 120 \ \Omega$$

$$L = \frac{X_L}{2\pi f}$$

$$= \frac{120}{2 \times 15.75}$$

$$= \frac{120}{98.91}$$

$$= 1.21 \ \text{H}$$

Capacitive Reactance

Capacitors affect the voltage timing in a circuit that contains this component. The respective waveforms for voltage and current in an ac capacitive circuit are shown in Figure 12-12. In this type of circuit, the current leads the voltage (remember the ICE man). The reaction to circuit changes due to capacitive values is called capacitive reactance. It is identified by the capital letter X and the subscript C (X_C). The formula for determining the exact amount of capacitive reactance in a circuit is

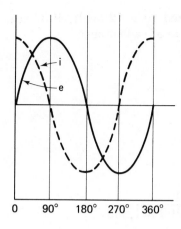

0   90°   180° 270° 360°

Figure 12-12. Voltage lags behind current in an inductive circuit.

$$X_C - \frac{1}{2\pi fC}$$

In this formula X_C is given in ohms, f is the frequency in hertz, and C is the capacitance in units of the farad. Use the following two examples to see how these values are determined.

Example 3. A 260-microfarad capacitor is wired to a 120-volt, 60-hertz source. Find the capacitive reactance and the current flow through the capacitor.

$$X_C = \frac{1}{2\pi fC}$$

$$= \frac{1}{2 \times 3.14 \times 60 \times 0.00026}$$

$$= \frac{1}{0.09797}$$

$$= 10.21 \ \Omega$$

$$I = \frac{E}{X_C}$$

$$= \frac{120}{10.21}$$

$$= 11.753 \ A$$

Example 4. A capacitor has a current flow of 100 milliamperes when connected to a 120-volt, 400-hertz source. Find the value of capacitance.

$$X_C = \frac{E}{I}$$

$$= \frac{120}{0.1}$$

$$= 1{,}200 \ \Omega$$

$$C = \frac{1}{2\pi f X_C}$$

$$= \frac{1}{2 \times 3.14 \times 400 \times 1{,}200}$$

$$= \frac{1}{3{,}024{,}400}$$

$$= 0.3 \ \mu\text{F}$$

IMPEDANCE IN AN AC CIRCUIT

The term resistance is used to represent an opposition to the flow of current in a circuit. Resistance is used in conjunction with Ohm's law. It normally is found in use with the description of action in a dc circuit. The values of voltage and current in the dc circuit are constant values once the circuit is functional. When an ac voltage is supplied to a circuit, there is a different set of conditions. This is due to the constantly changing value of the ac waveform.

Because of the continually changing voltage and current in the ac circuit, a different term is used to describe the opposition to current flow. This term is *impedance*. Impedance works in an ac circuit in a manner similar to resistance in a dc circuit. The letter used to represent impedance is the capital letter Z. Impedance is also measured in units of the ohm.

The values of voltage, current, and impedance are used in a manner similar to the previous use of Ohm's law. There is an application of Ohm's law for an ac circuit. The formula is very similar to that used for dc circuits except impedance is substituted for resistance. The formula reads

$$E = I \times Z$$

where E is voltage, I is current, and Z is the impedance in units of the ohm. The three values may be displayed in a triangular form much like the one used to display Ohm's law for dc circuits. It is shown in Figure 12-13. Solution for the unknown circuit value is done in the

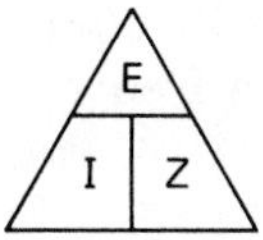

$$E = I \times Z$$

$$I = \frac{E}{Z}$$

$$Z = \frac{E}{I}$$

Figure 12-13. Ohm's law for AC circuits. Impedance (Z) is substituted for resistance.

same manner as with dc circuits.

There are some variations with the rules for finding impedance that apply to either a capacitive or an inductive circuit. These are required because of both the constantly changing values and the phase shift between voltage and current caused by either capacitance or inductance. The inclusion of either of these two components requires the addition of the values when solving for circuit values. The amount of reactance is a required part of the formula in either case. The formulas are

$$\text{Inductive circuit} - Z = \sqrt{R^2 + X_L{}^2}$$

$$\text{Capacitive circuit} - Z = \sqrt{R^2 + X_C{}^2}$$

Use of either of these formulas will provide an impedance value for an ac circuit. Examples of each are given.

Example 5. A series inductive circuit contains an inductance of 40 millihenries and 10 ohms of resistance. The frequency of the ac circuit is 60 hertz. The applied voltage is 120 volts ac. Find the reactance and the impedance of this circuit and the total current flow.

$$X_L = 2\pi f L$$

$$= 2 \times 3.14 \times 60 \times 0.04$$

$$= 15.07 \ \Omega$$

$$Z = \sqrt{R^2 + X_L^2}$$

$$= \sqrt{10^2 + 15^2}$$

$$= 18 \ \Omega$$

$$I = \frac{E}{Z}$$

$$= \frac{120}{18}$$

$$= 6.67 \ A$$

Example 6. A series capacitive circuit contains the following values: $C = 25$ microfarads, $R = 100$ ohms, and ac source voltage is 120 volts with a frequency of 400 hertz. Find the reactance, impedance, and current flow of the circuit.

$$X_C = \frac{1}{2\pi f C}$$

$$= \frac{1}{2 \times 3.14 \times 400 \times 0.000025}$$

$$= \frac{1}{0.0628}$$

$$= 15.92 \ \Omega$$

$$Z = \sqrt{R^2 + X_C^2}$$

$$= \sqrt{100^2 + 15.92^2}$$

$$= 101.26 \ \Omega$$

$$I = \frac{E}{Z}$$

$$= \frac{120}{101.26}$$

$$= 1.185 \ A$$

Many electrical circuits contain components of capacitance, inductance, and resistance. When this is the situation, the voltage across the capacitor is out of phase with the voltage across the resistance and inductance. At the same time that this is occurring, the current across the inductance is out of phase with the current through the resistance and capacitance. Capacitive reactance opposes inductive reactance in a circuit containing both components. The circuit value of impedance depends upon the difference between these two values. The formula for impedance in a series RLC circuit is

$$Z = \sqrt{R^2 + (X_L - X_C)^2}$$

Example 7. A series RLC circuit has the following values. $R = 50$ ohms, $X_C = 80$ ohms, $X_L = 100$ ohms. Find the impedance of the circuit and the current flow when 120 volts, 60 hertz is applied.

$$Z = \sqrt{R^2 + (X_L - X_C)^2}$$
$$= \sqrt{50^2 + (100 - 80)^2}$$
$$= 50 \ \Omega$$
$$I = \frac{E}{Z}$$
$$= \frac{120}{50}$$
$$= 2.4 \ A$$

RESONANCE

The voltage drops developed across each component in an ac series *RLC* circuit differ from those in a dc circuit. The reason for this is the phase shift that occurs in the ac circuit. When attempting to solve for circuit values in an ac circuit, one must take into consideration the phase angle of voltage across each component. There is a point, however, when the capacitive reactance and the inductive reactance cancel each other. This occurs when the two reactances are equal. When these conditions occur, the circuit is said to be at *resonance*. The specific frequency at which resonance occurs is called the *resonant frequency* of the circuit.

If the formula for impedance is used and the values of capacitive and inductive reactance are equal, all that is left in the circuit to oppose current flow is the resistance value. Apply this statement to a series resonant circuit. The only component left to establish the amount of current flow is the resistance. A low resistance value will permit a large current flow. A large value will greatly limit the current flow in the series resonant circuit. This is shown in Figure 12-14.

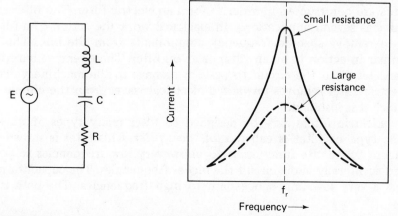

Figure 12–14. At resonance the series *LRC* circuit has a low impendance. A large current flow develops through the components.

Changing the value of capacitance or inductance in the circuit will change the resonant frequency. When this occurs, current flow in the circuit will be reduced to a lower level. This is illustrated by the graph.

A parallel resonant circuit has different characteristics than does its series counterpart. In this type of circuit, as illustrated in Figure 12-15, the opposition is to voltage rather than current. The voltage

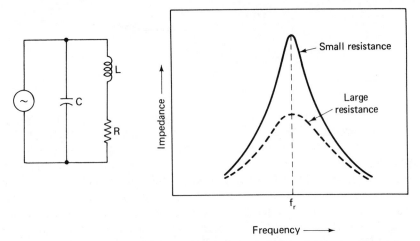

Figure 12-15. At resonance the parallel *LRC* circuit has a high impedance. A large voltage drop develops across the components.

drops developed across each inductive and capacitive component often are quite large. These voltage drops are 180 degrees out of phase with each other. This tends to cancel the current flow from the source. It also produces a very high impedance in the circuit and a very large voltage drop across the resistive component.

Resonance is useful in electrical circuit applications. Resonant circuits are constructed in order to form an electric filter. The filter circuit has several applications. In electrical work the action of a filter is to remove specific frequency components from the line. This is similar in action to an air filter or an oil filter. These remove harmful particles from the circuit to prevent damage to the machinery. The electric filter removes unwanted electrical waves from the circuit in which it is installed.

Electric filters may be designed to filter many types of waves. One type of filter is called a *low pass filter*. This filter is shown in Figure 12-16. Its function is to allow very low frequencies to pass and to literally short circuit the higher frequencies. The capacitor offers a very low impedance path to high frequencies. The path is a

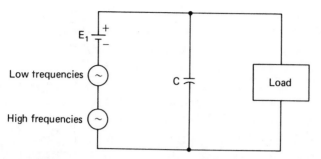

Figure 12-16. Electronic filters are designed to either pass or reject specific frequency ranges. This low pass filter rejects high frequencies and allows low frequencies to pass.

much lower impedance than the load impedance at high frequencies. These frequencies bypass the load by passing through the capacitor. Another type of filter is shown in Figure 12-17. This is called a high

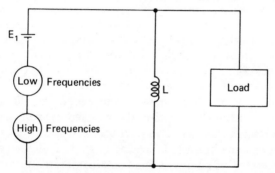

Figure 12-17. This is a high pass filter circuit. All low frequencies are passed around the load by the inductor and return to circuit common.

pass filter circuit. The inductance offers a low impedance to high frequencies. These bypass the load in the same manner as discussed for the low pass filter.

A third type of filter is called a bandpass filter. This filter is shown in Figure 12-18. It has a series resonant circuit and a parallel resonant circuit. Both of these circuits are connected to the load. The series resonant circuit is designed to pass a specific frequency. The size of the components used as the filter will determine the frequency of resonance. The parallel resonant circuit provides a high impedance at the resonant frequency. It also offers a low impedance to all other frequencies. This parallel resonant circuit will bypass all frequencies other than at resonance. The series resonant circuit bypasses all frequencies other than the resonant frequency. The values of inductance and capacitance used for each filter determine the resonant frequency

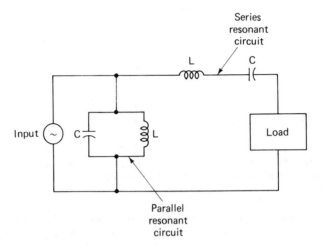

Figure 12-18. The bandpass filter rejects both high and low frequencies. It is designed to pass a specific range only.

of the circuit. This type of filter often is designed to operate at a specific frequency in the middle of some group of frequencies. It is called a *bandpass* filter because it allows a specific band of frequencies to pass and rejects all other frequencies in the system.

Another major use of resonant circuits is to transfer energy from one circuit to another. These circuits, as used in radio and television sets, are called *tuned* circuits. The purpose of these circuits is to transfer energy from one circuit to a second circuit. The device often used to effect a transfer of energy is the transformer. The transformer has at least two identified sets of windings. Energy is transferred from one winding to a second winding at great efficiency. In most cases the two transformer windings are not connected to each other. Chapter 13 discusses transformer action in detail.

SUMMARY

All electrical components exhibit some properties of inductance and capacitance. In some cases the amounts are minimal and are not related to circuit action. In other cases either of these properties will affect circuit action. The most often encountered effect is a shift in timing, or phase, between the voltage and current in the circuit. Phase shift relates to the starting time of the electrical waves. Waves in phase with each other start at the same time. Waves out of phase have a different starting time. Phase relationships are described by relating to the number of degrees of rotation of a generator cycle.

Purely resistive circuits have voltage and current across the load in phase with each other. Circuits that contain capacitance and resistance have the current leading the voltage. Voltage will lead current in an inductive circuit.

In a circuit that contains a capacitor there is a time lag for the voltage to rise to its full value. The time lag is called the time constant. Voltage rises 63.2% for one time period. The time period is also called the time constant. Five time constants are required for a full rise. The same holds true when the voltage falls.

Inductive circuit current reacts in the same manner as does voltage in the capacitive circuit. Five time constants are required for either a full rise or a full fall of current. This is true in any circuit where either the voltage or the current changes value.

An ac circuit containing inductance is said to be reactive. The inductance is called inductive reactance (X_L). A circuit used with ac and containing a capacitance is also reactive. The circuit is called a capacitive reactive circuit (X_C).

Impedance is the term used to describe resistance in an ac circuit. This is due to the constantly changing values as the voltage and current rise and fall. The letter Z is used to represent impedance. It may be substituted for resistance and used in Ohm's law for ac circuits.

Circuits that contain inductance, capacitance, and resistance have a point of resonance. This frequency is dependent upon the electrical size of the components. Some resonant circuits are used as filters to remove unwanted voltages. Other filter circuits are designed to accept a frequency or band of frequencies and to reject all other frequencies. These are called bandpass filters.

Resonant circuits are also used to transfer energy from one circuit to a second circuit. These circuits are used in radio and television sets. Transformers also use this principle to transfer power.

FORMULAS

RL charge/discharge: $t = 5\dfrac{L}{R}$

RC charge/discharge: $t = 5RC$

RL time constant: $t = RC$

RC time constant: $t = \dfrac{L}{R}$

Inductive reactance: $X_L = 2\pi fL$

Capacitive reactance: $X_C = \dfrac{1}{2\pi fC}$

Impedance: $Z = \dfrac{E}{I}$

$\qquad = \sqrt{R^2 + X_L^2}$ (inductive)

$\qquad = \sqrt{R^2 + X_C^2}$ (capacitive)

Resonance: $X_L = X_C$ or $f_o = \dfrac{1}{2\pi\sqrt{LC}}$

PROBLEMS

12-1. Find the charge time for a 100-μF capacitor series connected to a 100-kΩ resistor.

12-2. Find the charge time for a 0.05-μF capacitor series connected to a 1.0-MΩ resistor.

12.3. Find the charge time for a 10-H inductor series connected to a 500-kΩ resistance.

12-4. Find the charge time for a 500-mH inductor series connected to a 1-kΩ resistor.

12-5. Find the impedance of a series circuit containing 1,500 Ω of resistance and 1,000 Ω of inductive reactance.

12-6. Find the impedance of a series circuit containing the following: $R = 100\ \Omega, X_C = 25\ \Omega, X_L = 50\ \Omega$.

12-7. A 25-H inductor is series connected to a 120-V, 60-Hz source. Find the reactance and the ac current.

12-8. A 0.5-H inductor is series connected to a 15,750-Hz source. Find the reactance and the current if the voltage is 500 V.

12-9. A 100-μF capacitor is series connected to a 120-V, 60-Hz source. Find the reactance and the current.

12-10. A 120-μF capacitor is series connected to a 500-Hz, 12-V source. Find the reactance and the current.

QUESTIONS

12-1. Opposition to changing circuit values is called

    a. ohmic resistance

    b. impedance

    c. reactance

    d. reluctance

12-2. The relation of any two or more electrical waves starting times is called the

    a. phase relation

      b. starting relation

      c. cycle timing

      d. none of the above

12-3. Starting time relationship of electrical waves is based upon a

      a. 180-degree circular motion of a magnetic field

      b. 360-degree circular motion of a generator

      c. general description of wave shape

      d. size relationship of the electrical waves

12-4. Starting time relationship of two waves in a resistive circuit is

      a. equal

      b. 90 degrees apart

      c. 180 degrees apart

      d. 270 degrees apart

12-5. In an inductive circuit

      a. voltage leads current

      b. current and voltage start together

      c. current leads voltage

      d. none of the above

12-6. In a capacitive circuit

      a. voltage leads current

      b. current and voltage start together

      c. current leads voltage

      d. none of the above

12-7. Time for a capacitor or an inductor to charge 63.2% is called its

      a. charge time

      b. timing charge

      c. time constant

      d. time holdback constant

12-8. The number of units required for a full charge is

      a. 1

      b. 2

      c. 3

      d. 5

12-9. Opposition to current in an ac circuit is called

      a. ohmic holdback

   b. ohmic value

   c. reactance

   d. impedance

12-10. When capacitive and inductive units are equal, the circuit is

   a. reactive

   b. low capacitive

   c. noninductive

   d. resonant

13

Conductors and Transformers

The creation of electric energy is rather meaningless unless this energy can be utilized. There are very few instances where both the generator and the load are found in the same geographic location. In most cases the electric energy has to be transported from the source to the load. The transportation of power is accomplished by using wires. The wires act as conductive paths for the power. The size, length, and composition of these wire conductors is of great importance to the electrical technician.

WIRES AND CONDUCTORS

Many materials are classified as conductors. Each of these materials has the capability of giving up free electrons from its atomic structure. The ease with which the conductor gives up these electrons is directly related to the conductor's ability to carry an electric current. Conductive materials include silver, gold, copper, and aluminum. Each material has a specific resistance rating. These, and other conductors, are shown in Figure 13-1. This chart lists common conductors by their specific resistance. Silver has the least resistance on this chart. This is followed by copper. Copper, being a more abundant material than silver, has been utilized as the most common wire conductor. Some applications use aluminum conductors. These applications are limited to high-power circuits. They are seldom found in the home. Each conductor listed in the chart has an ohmic resistance value. This is figured on the basis of a sample 1 foot long and 1 circu-

Specific Resistance
of Conducting Materials

Material	ρ Approximate Specific Resistance (Ω/cm ft) at 20°C (68°F)
Silver	9.8
Copper	10.4
Gold	14.7
Aluminum	17.0
Tungsten	33.2
Nickel	50.0
Iron	58.0
Steel	96.0
Nichrome	660.0
Carbon	20,000 (approx.)

Figure 13-1. Conductivity ratings for common current-carrying conductors.

lar mil in diameter. One circular mil is defined as the cross-sectional area of a round wire having a diameter of 1 mil (0.001 in.). The circular mil area of a conductor is equal to the square of the diameter in mils. Thus, doubling the circular mil area of a conductor increases the circular mil area by four. The cross-sectional area of a conductor is very important in relation to the amount of current the conductor can safely handle.

Information relating to the size and other characteristics of copper wire is found in a copper wire table. One such table is shown in Figure 13-2. This table contains a wealth of information. The first column gives standard wire sizes. The sizes relate to the gage number of the wire. The numbers used to indicate the gage number are inversely proportional to the size of the wire. Large-numbered wires have the smallest size.

Columns 2 and 3 on the chart show the diameter and cross-sectional area of the wires. The largest-sized wires have the smallest numbers. Column 4 shows the resistance of the wire per thousand feet. The resistance is also inversely proportional to the number. Larger wires have less resistance than do their smaller counterparts. Less resistance means that the wire is able to handle larger amounts of current. Column 5 shows the weight of copper wire per one thousand feet.

The last three columns show the amount of current that can safely be carried by each size of wire. The type of insulating material affects the current-carrying capacity, also. This factor is important when selecting the proper insulation for a specific application. Most conductors used to wire machinery or in the home use a plastic

Standard Annealed Copper Wire Solid American wire gage (Brown and Sharpe)

1	2	3	4	5	6	7	8
			Resistance, Ω/1000 ft		Allowable current capacity, A		
Gage number	Diameter, mils	Area, cir mils	25°C (77°F)	Weight lb/1,000 ft	Rubber insulation	Varnished cambric insulation	Other insulations
0000	460.0	211,600.0	0.0500	641.0	225	270	325
000	410.0	167,800.0	0.0630	508.0	175	210	275
00	365.0	133,100.0	0.0795	403.0	150	180	225
0	325.0	105,500.0	0.100	319.0	125	150	200
1	289.0	83,690.0	0.126	253.0	100	120	150
2	258.0	66,370.0	0.159	201.0	90	110	125
3	229.0	52,640.0	0.201	159.0	80	95	100
4	204.0	41,740.0	0.253	126.0	70	85	90
5	182.0	33,100.0	0.319	100.0	55	65	80
6	162.0	26,250.0	0.403	79.5	50	60	70
7	144.0	20,820.0	0.508	63.0			
8	128.0	16,510.0	0.641	50.0	35	40	50
9	114.0	13,090.0	0.808	39.6			
10	102.0	10,380.0	1.02	31.4	25	30	30
11	91.0	8,234.0	1.28	24.9			
12	81.0	6,530.0	1.62	19.8	20	25	25
13	72.0	5,178.0	2.04	15.7			
14	64.0	4,107.0	2.58	12.4	15	18	20
15	57.0	3,257.0	3.25	9.86			
16	51.0	2,583.0	4.09	7.82	6		
17	45.0	2,048.0	5.16	6.20			
18	40.0	1,624.0	6.51	4.92	3		
19	36.0	1,288.0	8.21	3.90			
20	32.0	1,022.0	10.4	3.09			
21	28.5	810.0	13.1	2.45			
22	25.3	642.0	16.5	1.95			
23	22.6	509.0	20.8	1.54			
24	20.1	404.0	26.2	1.22			
25	17.9	320.0	33.0	0.970			
26	15.9	254.0	41.6	0.769			
27	14.2	202.0	52.5	0.610			
28	12.6	160.0	66.2	0.484			
29	11.3	127.0	83.4	0.384			
30	10.0	100.0	105.0	0.304			
31	8.9	79.7	133.0	0.241			
32	8.0	63.2	167.0	0.191			
33	7.1	50.1	211.0	0.152			
34	6.3	39.8	266.0	0.120			
35	5.6	31.5	335.0	0.0954			
36	5.0	25.0	423.0	0.0757			
37	4.5	19.8	533.0	0.0600			
38	4.0	15.7	673.0	0.0476			
39	3.5	12.5	848.0	0.0377			
40	3.1	9.9	1070.0	0.0299			

Figure 13-2. A copper wire table provides information relating to size, capacity, and ohmic value of the wire.

insulation. The ratings for this type of insulation are found in column 8.

The information found in a copper wire table is useful to the electrical technician. One way of using these data is to determine the size of conductors required when wiring a machine control box. Wire size and current-carrying capability are needed when connecting a machine or motor to the source. At other times the information relating to wire size is used to calculate the number of wires required for a solenoid or transformer.

A major obstacle to the flow of an electric current is the heat buildup in the conductor. Heat is developed owing to the movement of electrons through the wire. A current of larger value than the wire is capable of handling will create heat on the wire. This heat tends to increase the resistance of the wire. The added resistance in the wire will reduce the total current flow in the circuit. Under normal conditions the resistance of power wires is a very low value. Unless the length of the wire is very long, the wire resistance is not significant.

The resistance of the wire may be a significant factor under certain circumstances. An example of this is illustrated in Figure 13-3.

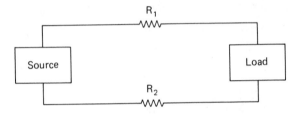

Figure 13-3.   Wire resistance, as described in the text, will affect current flow in the circuit.

Resistors R_1 and R_2 represent the wire resistance in this circuit. Values of 120 volts for the source and 4 ohms of resistance for the load are given as typical values. Using Ohm's law, the circuit current when R_1 and R_2 are not considered is

$$I = \frac{E}{R} = \frac{120}{4} = 30 \text{ A}$$

Now consider what happens as each of the wire resistances is increased to 1 ohm. The total resistance in the circuit increases from 4 to 6 ohms. Circuit current is now

$$I = \frac{E}{R} = \frac{120}{6} = 20 \text{ A}$$

Circuit current drops 10 A owing to the increase in resistance in the wires.

Consider what happens to the operating voltage across the load when R_1 and R_2 are not considered. All the 120 volts developed at the source are applied to the load. The power at the load is

$$P = E \times I = 120 \times 30 = 3{,}600 \text{ W}$$

When R_1 and R_2 increase to 1 ohm each, there is a voltage drop developed across each of the wires. The voltage drop reduces the voltage available at the load. Using the information developed earlier, there is a total current flow of 20 A as the wires increase in resistance. The voltage drop across the load is figured using this lower current:

$$E_{\text{Load}} = I \times R = 20 \times 4 = 80 \text{ V}$$

The power available at the load is also reduced under these conditions.

$$P = E \times I = 80 \times 20 = 1{,}600 \text{ W}$$

Power is directly related to the ability to perform some work. A reduction in voltage across the load has resulted in a power loss of 55%. This type of situation is beyond any accepted standards. The solution is to use a wire size that has the capability of handling the full 30 amperes of current without any significant wire resistance losses.

The addition of 1 or 2 ohms of resistance into the circuit does not seem significant. Actually, the opposite is true. All resistances affect the circuit action. The resistance developed by a defective or oxidized wire lug may cause the development of the same problem. This is particularly true in circuits that operate at low voltage but require very high currents. Typical applications include forklift vehicles and the automobile. Consider the results of this circuit as an unwanted resistance develops. The original circuit has a 12-volt source. The load uses 40 amperes of current. The load resistance is

$$R = \frac{E}{I} = \frac{12}{40} = 0.3 \ \Omega$$

The addition of 0.2 ohms of resistance due to an oxidized connector drops the current to the load to 24 amperes. This drops the operating power from 480 to 288 watts. This is a 40% loss in operating power. This loss, if applied to the cranking circuit of an automobile engine, would probably keep the engine from cranking fast enough to start. The transfer of power from the source to the load is of great importance. The next section discusses this action and its importance.

MAXIMUM POWER TRANSFER

The previous section describes what happens when unwanted circuit resistances occur. This information relates to the transfer of electrical power from the source to the load. A circuit that is used to illustrate power transfer action is shown in Figure 13-4. Maximum

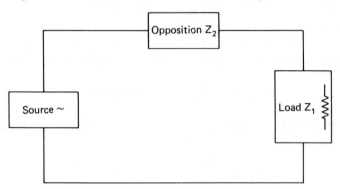

Figure 13-4. Power transfer between source and load must include all opposition to this transfer.

power transfer in a circuit occurs when the source impedance (or resistance) is equal to the load impedance. This is illustrated by assigning values to each component in the sample circuit. Use a source voltage of 120 volts. The box marked "opposition" is given an arbitrary value of 10 ohms. Call this an impedance in order to conform to ac circuit descriptions. It is identified as Z_2 in this circuit. This value represents the impedance of the source. The load is identified as impedance 1 (Z_1). It has an impedance of 1 ohm. Values of current can be determined by the use of Ohm's law for ac circuits:

$$I = \frac{E}{Z} = \frac{120}{11} = 10.9 \text{ A}$$

The voltage drop developed across the load is

$$E = I \times Z = 10.9 \times 1 = 10.9 \text{ V}$$

The power developed by the load is

$$P = E \times I = 10.9 \times 10.9 = 118.8 \text{ W}$$

The values of voltage, current, and power for a variety of load impedances are given in Figure 13-5. To produce the results shown, the source voltage is maintained at 120 volts. The data in the table show the current, voltage drop, and power consumption for the load with

E_{source} volts	Z_2, ohms	Z_1, ohms	I, amperes	E_{Z1}, volts	P_{Z1}, watts
120	10	1	10.91	10.91	119
120	10	5	8	40	320
120	10	10	6	60	360
120	10	15	4.8	72	345
120	10	20	4	80	320
120	10	100	1.09	109.0	118.8

Figure 13-5. Relationship of circuit values. Maximum power transfer occurs when the impedances of the load and the source are the same.

different load impedances. Note that the load power is greatest when the impedance of the load is equal to the impedance of the source. The conditions given for the rest of the load impedances show a loss of power at the load. This power loss is developed in the wires or as a part of an electrical wave that opposes the applied voltage in the circuit. This reflected wave is developed by the mismatched impedances. It reduces the operating efficiency of the load in the circuit.

THE TRANSFORMER

If you have an inquisitive mind, you might be asking about how to transfer maximum power when the source and the load impedances do not match each other. The solution to this problem is to use an impedance-matching device. This device is connected between the source and the load in the circuit. This is illustrated in Figure 13-6.

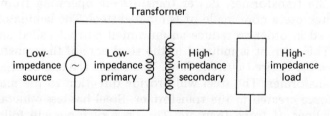

Transformer

Low-impedance source — Low-impedance primary — High-impedance secondary — High-impedance load

Figure 13-6. A transformer is used to transfer power when impedances are not matching.

One type of device inserted in the circuit that has the capability to match impedances is called a transformer. The basic transformer has three parts, the primary winding, the secondary winding, and the core. These are shown in Figure 13-7.

The transformer works on the principle of electromagnetic induc-

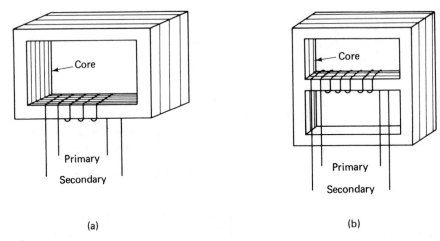

Figure 13-7.   Construction of a transformer showing windings and the core.

tion. A varying voltage in the primary winding, or primary, will establish a magnetic field around the wires. The magnetic field rises as the current in the circuit increases. The current increases owing to a rising voltage value in the circuit. The magnetic field falls as the current decreases in the circuit. This constantly varying magnetic field is typical in an ac circuit. The number of turns of wire and the size of the wire on the primary winding produce a measurable impedance. This impedance matches the impedance of the source. In most instances the impedance of the primary winding of a transformer used in power circuits is very low. It is most likely on the order of 10 ohms.

The second portion of the transformer is the core. The material used for the core is dependent upon the frequency of the voltage applied to the transformer. Power transformers operating from a 60-hertz source use a core made of laminated steel. The laminated steel core is used in order to reduce an unwanted current called an *eddy* current. This current is induced in the steel core of the transformer. It opposes the desired current flow. The result is a loss of efficiency in the transformer. The steel will provide direction to the magnetic lines of force created in the transformer. Steel has less reluctance to magnetic lines of force than air. The lines of force will follow the path of least reluctance. The result is a strong electromagnetic field. The strong magnetic field is required if maximum power is to be transferred from the primary winding to the secondary winding of the transformer.

The secondary winding of the transformer is designed to match the impedance of the load. This winding is wound either on top of or next to the primary winding of the transformer. The wires of both windings parallel each other. Lines of force from the primary move at right angles to the wire. Maximum voltage is induced on the secondary winding when it is at right angles to the lines of force. Winding both of the windings so that their respective wires are parallel to each other sets the conditions for maximum transfer of energy. The voltage induced in the secondary winding causes a current to flow. This is true only when a load is connected to the secondary. The load completes the circuit on the secondary winding.

Load and No-Load Conditions

Assume that the circuit shown in Figure 13-6 does not have a load connected to the secondary winding. There is a source voltage connected to the primary winding. Current flows in the primary of the transformer. Lines of force induce a voltage on the secondary winding. The unloaded secondary will act as a source under these conditions. The induced voltage at the secondary is of an opposite polarity to that of the primary. This is due to the motion of the lines of force that created the secondary voltage. The secondary voltage has its own magnetic field. This field is opposite the polarity of the primary winding. It cuts across the primary windings. The two fields add their respective forces. In effect, the forces tend to cancel each other. The result is a very low current flow in the primary winding. The power used in this situation is only the power required to magnetize the core of the transformer.

Now assume that a load is connected to the secondary winding. Current will flow from the winding through the load. This action tends to reduce the reverse magnetic field established in the primary. The result is a larger current flow in the primary of the transformer. More power will be transferred from the primary to the secondary winding.

Changing the impedance at the load connected to the secondary winding will change the current flow in the primary winding. Primary current will increase as the secondary load current increases. Primary current will decrease as the load impedance is made larger and less current flows in the secondary winding. The transformer will automatically adjust to changes in secondary current in this manner. There is a point of no return in this system. When the load current exceeds the capabilities of the primary, the transformer heats and may often melt the wires in the primary. This results in an open-circuited primary winding and zero output voltage at the secondary.

Step-up and Step-down Transformers

Transformers often serve a function in addition to impedance matching. This function is to provide a level of operating voltage other than that furnished by the source. The type of transformer described so far is able to maintain the amplitude of source voltage at the secondary of the transformer. There are many instances where the secondary voltage needs to be of a different value. In some of these situations the secondary voltage requirements are for a voltage of less amplitude than the primary voltage. Transformers that have the capability of reducing the source voltage are called step-down transformers.

Assume that our model transformer has an input voltage from the source of 120 volts. Assume also that there are 120 turns of wire on the primary winding. The transformer is connected to its source, and no load is connected to the secondary. The magnetic field created in the primary as the current flows produces an EMF with the opposite polarity as the source. The 120 turns of wire connected to a 120-volt source will produce an opposite polarity voltage of 120 volts. This indicates that each turn of wire on the primary winding has 1 volt induced on it.

A secondary winding with 120 turns of wire is connected to its load. The magnetic forces from the primary will induce a voltage equal to 1 volt for each turn of wire on the secondary. If the secondary has 120 turns of wire, the induced voltage at the secondary is 120 volts. If the number of turns of wire on the secondary is reduced to 24, the voltage developed on the secondary is 24 volts. Step-down transformers are made by changing the ratio of turns of wire between the primary winding and the secondary winding.

An additional feature of this procedure is related to the currents in the primary and the currents in the secondary. The transformer is not a 100% efficient unit. We can assume that its efficiency is high, however. For purposes of discussion, assume the efficiency to be 100%. Our model transformer has a 120-volt rating on its primary winding. The primary current is 0.5 amperes. The power consumed by the primary is 60 watts. The power in the primary is transferred to the secondary if the transformer has a 100% efficiency. The secondary also has a power consumption of 60 watts if this is true. The secondary voltage developed by the reduced number of turns of wire is 24 volts. The power in the secondary is 60 watts. The current in the secondary that will produce 60 watts is 2.5 amperes. When the power transfer is maintained, a step-down transformer actually produces a lower voltage and a higher current in its secondary. In a practical transformer the secondary current is determined by the load. The size of wire used to wind the secondary limits its current capacity.

Transformers use this same principle of ratio of turns between primary and secondary to step up a voltage. In a step-up transformer the secondary has more turns of wire than used on the primary. The power in the primary is still equal to the power in the secondary. Using Watt's law, if the voltage increases and the power is held at a constant value, the current will decrease. The exact values depend upon the load and the number of turns of wire.

A mathematical relationship may be established between the ratio of turns and the ratio of voltages. This relationship is

$$\frac{E_p}{E_s} = \frac{N_p}{N_s}$$

where E_p is the voltage in the primary, E_s is the voltage in the secondary, N_p is the number of turns of wire in the primary, and N_s is the number of turns of wire in the secondary.

Transformers may have multiple windings. These multiple windings may be found in either the primary or the secondary of the transformer. There are some instances where both the primary and the secondary have multiple windings. Each winding in a transformer has a polarity rating. This can be seen if time could be stopped for an instant. This is illustrated in Figure 13-8. The graphic symbol for an

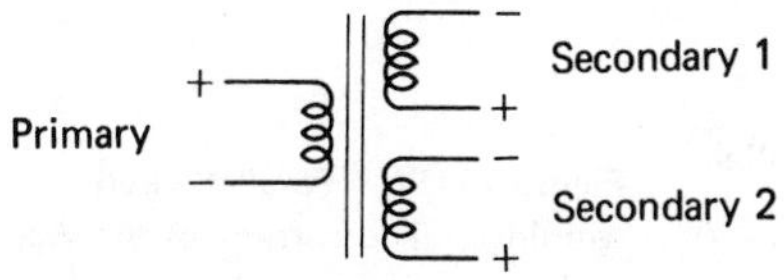

Figure 13-8. Polarities for primary and secondary windings at one moment in time.

iron-cored transformer is shown. This transformer has one primary and two secondary windings. The current flow in the transformer is stopped for one instant. The polarities of the primary and secondary windings are shown at this moment in time. A polarity reversal occurs owing to the movement of the magnetic field in the transformer.

In many cases, transformers are produced with more than one secondary winding. The windings may be wired together to take advantage of this. Secondary windings may be series wired to increase the secondary voltage. The voltages developed across each winding will add. This is shown in Figure 13-9. This supports Kirchhoff's law

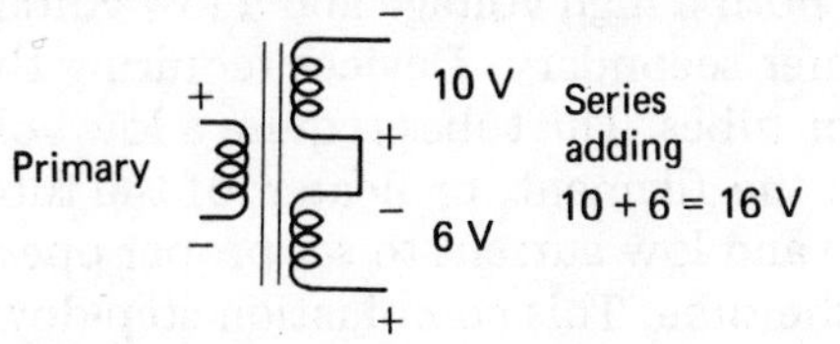

Figure 13-9. Windings wired in a series-adding form. The total voltage is the sum of the voltage developed in each of the windings.

for voltages in a closed loop. When a load is connected to the two sources, the total source voltage is equal to the sum of each voltage.

It is often difficult to tell which wire lead on the secondary is the start of the winding and which is the end of the winding. The transformer secondary windings may be connected in reverse polarity. This is shown in Figure 13-10. Kirchhoff's law still applies in this situ-

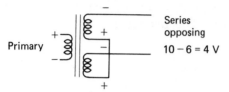

Figure 13-10. Series-opposing wiring connection. The output voltage is still the sum of the two voltages.

ation. The result is the sum of two opposite polarity voltages. This actually develops a voltage value that is equal to the difference between the two voltages. When each winding produces the same value of voltage, they cancel each other. The result is a zero voltage across the two windings. This is not a desirable situation and should be avoided.

There are instances when two secondary windings are parallel connected. This only works successfully if both windings produce the same voltage. This is shown in Figure 13-11. When the windings are

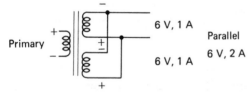

Figure 13-11. Parallel-wired windings. The current of the secondary is the sum of each winding's current capacity.

wired in parallel, the current of each is added to produce a total current. This follows Kirchhoff's current law. The currents entering a junction equal the current leaving a junction. The 1-ampere currents from each winding add at the junction to produce a 2-ampere current to go to the load.

In some other cases a transformer may have two primary windings. Each winding is usually designed to operate from a 120-volt source. The windings are parallel wired for 120-volt operation. They are series wired when the transformer is wired to a 240-volt source. There are instances, also, when both a high voltage and a low voltage are required from the transformer secondary. Devices requiring this combination usually use vacuum tubes. The tubes require a low voltage and high current to operate the filament, or heater, of the tube. They also require a high voltage and low current to set proper operating conditions for the rest of the tube. This combination step-down and step-up transformer may even have more than one pair of secondary windings. The actual number of windings becomes a design factor. A typical power transformer schematic drawing is shown in Figure 13-12.

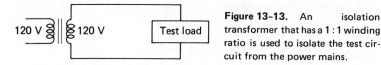

Figure 13-12. A power transformer often has a high voltage-low current winding and a low voltage-high current winding.

One other consideration is the isolation transformer. It is used to isolate the primary circuit from the secondary circuit. These transformers have a 1:1 ratio of windings and voltages between the primary and the secondary. Their specific purpose is to provide electrical isolation of the secondary circuit. One application for this transformer is on the test or repair bench. The circuit or device being tested is connected to the transformer secondary winding. The primary winding is connected to the power source. There is no direct connection between primary and secondary windings. The resulting isolation of these circuits is related to safety. Isolation from the power source often protects the circuit and its operator from the danger of electrical shock or equipment damage if a wiring error or short circuit occurs. A schematic for an isolation circuit is shown in Figure 13-13.

120 V ⧑‖⧑ 120 V | Test load |

Figure 13-13. An isolation transformer that has a 1 : 1 winding ratio is used to isolate the test circuit from the power mains.

TRANSFORMERS IN DC CIRCUITS

The theory used for transformer action shows the inducement of voltage on a wire only when the magnetic field is changing. A stationary magnetic field in one wire will not induce a voltage on adjacent wires. This means that the transformer does not work in a dc circuit. The only time there is a change in the magnetic field's size in a dc circuit occurs when the circuit is either turned on or turned off. During this moment in time the field will expand to the level permitted by Ohm's law, or it will collapse to a zero level. This theory is used in practical dc circuits. Mechanical devices are available that will cause the dc current to rise and fall at some regular rate. These devices are often used in the automobile. The ignition points operate in such a manner. These are illustrated in Figure 13-14. The points are controlled by a cam. The cam is driven by the engine of the automobile.

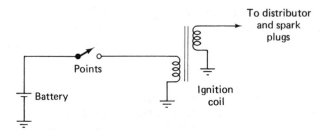

Figure 13-14.   A circuit where DC is "chopped" or interrupted in order to act as an AC value.

The points are in the primary circuit of the ignition coil. The current flow through the primary circuit is interrupted at a regular rate. This causes a rise and fall in current in the coil primary.

The secondary of the coil is connected to the distributor rotor and spark-plug circuit. Spark plugs receive the high voltage from the secondary in order to ignite the gases in the cylinder. The points are found in the distributor housing. This is a convenient mechanical arrangement for the manufacturer of the automobile.

Another device used to interrupt dc current is called a chopper or a vibrator. This device works on electromechanical principles. One such device's schematic diagram is shown in Figure 13-15. When the

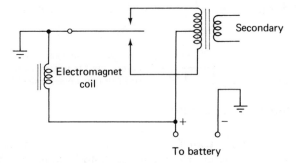

Figure 13-15.   A device called a vibrator is often used to provide an alternating DC. This is used with a transformer to provide high DC voltages.

circuit is first turned on, the magnet coil draws the points toward one set of contacts. This produces a current flow through one part of the primary of the transformer. The current is interrupted by the movement of the vibrating reed contact points and flows through the other half of the primary winding. The vibrating reed continually makes and breaks the circuit connection. This produces a pulsating wave through the action of the varying magnetic field. The transformer will accept this varying field and produce a voltage at its secondary windings.

SUMMARY

Electric generating plants are often separated geographically from the consumer of the power. It is necessary to use a conductor of electrical power to connect the source and the load. Wires made of copper or aluminum are used for this purpose. The size of a conductor is related to its ability to carry an electrical current. The cross-sectional area of the conductor has a gage number. The numbers relate to the size of the conductor. Gage numbers are in inverse proportion to the diameter of the wire. Conductors that have a low gage number are relatively large in size. They can handle large amounts of electric current.

Wire resistance generates heat in the wire as current flows. The resistance of the wire can be of great significance. A small resistance in the wire may often produce an undesired voltage drop between the source and the load. The result is a reduced amount of power at the load. Reduced power at the load means that less work can be accomplished. The consumer pays for all the power delivered to the load. Power losses result in a greater cost of operation.

Maximum power transfer is required in an efficient system. Maximum power transfer occurs when the source impedance is equal to the impedance of the load. When the source and load impedances do not match, an impedance-matching device is required. One such device is the transformer. A basic transformer has a primary winding, a secondary winding, and a core. The number of turns of wire on each winding is directly related to the voltage on each winding. The size of each wire determines the current capacity of the winding.

Transformers work on the principle of electromagnetic induction. A magnetic field is produced by the current flow in the primary. The magnetic field moves across the secondary winding wires and induces a voltage on these wires. The core, made of laminated steel, helps in concentrating the lines of force around the windings of the transformer.

Transformer winding ratios determine the voltage developed in the secondary winding. The transformer is used to step up or to step down voltage, as required. Power in the primary of the transformer is equal to power in the secondary. This allows the use of a low-voltage and high-current secondary. In some instances the transformer has both types of windings. In any case, the power in the secondary cannot exceed the power in the primary.

Transformers only react to the changing magnetic field produced as current rises and falls in the windings. Under normal conditions, transformers cannot operate on dc voltage. Special devices are used to turn the dc voltage on and off. The dc voltage then acts in a manner similar to an ac voltage. The result is a varying amplitude current that will make the transformer work.

QUESTIONS

13-1. Wire numbers and wire sizes ____ to each other

    a. are directly proportional

    b. are inversely proportional

    c. are not related

    d. have no sequence

13-2. Wire current capacity increases as

    a. wire size decreases

    b. wire length decreases

    c. wire size increases

    d. resistance increases

13-3. Opposition to current flow produces

    a. heat

    b. voltage drop

    c. current drop

    d. all of the above

13-4. When source impedance matches load impedance in a circuit

    a. minimum power transfer occurs

    b. energy creates heat

    c. maximum power transfer occurs

    d. voltage drops produce losses

13-5. A device used to match different impedances is called

    a. the transformer

    b. an impedance matcher

    c. an impedance adjustor

    d. a matchmaker

13-6. The voltage relationship between primary and secondary windings of a transformer depends upon

    a. wire size

    b. winding ratios

      c. applied voltage

      d. all of the above

13-7. Transformer action occurs when

      a. current rises

      b. current is constantly changing

      c. current is steady

      d. no current flows

13-8. Under normal conditions, transformers work with

      a. direct current

      b. alternating current

      c. both of the above

      d. neither of the above

13-9. Transformers function at

      a. 25-hertz frequencies

      b. 60-hertz frequencies

      c. all frequencies

      d. both a and b above

13-10. Transformers may have

      a. secondary step-down windings

      b. secondary step-up windings

      c. a variety of secondary windings

      d. none of the above

14

Resistive and Inductive Loads

The material found to this point in the book is related to basic electrical theories. These theories are applied in practice. The result of the application of the theories is that some work is accomplished. The term used in electrical descriptions of work is *load*. The load represents a device that will change electrical energy into a form that results in the performance of work. Light bulbs are loads. They transform electrical energy into light energy. Machines perform work. The machine is more often considered a mechanical device. It does, however, require a driving force to produce mechanical energy. In most cases the driving force is an electric motor.

Machines often have parts other than motors that require electrical energy. Some machines operate best on a dc source rather than an ac source. The ac from the source is changed into dc at the machine. The parts required to do this and their operation are required knowledge for persons working with electricity. This chapter presents material related to loads in electrical circuits. A variety of typical loads are discussed.

ELECTRICAL LOADS

Electric loads may be classified as either resistive or inductive in operation. This does not mean that some loads may be both resistive and inductive. It does mean that the major influence on the source is either resistive or inductive in nature. The descriptions in the following section are general in nature. There are additional loads that will fall into one of these two categories. The theories that are presented apply to all loads. Once the classification system is understood, the theory of operation is easy to apply.

RESISTIVE LOADS

The opposition to the flow of an electric current is called resistance. Resistance in any conductor is determined by the physical properties of the conductor. When an electric current flows through any type of conductor, the movement of the electrons produces a certain amount of heat. This is due to the electrical friction of the conductor.

When a conductor is selected, a major consideration is its ability to carry the electrical current with a minimum of heat production. Cost is also a consideration. In most cases either copper or aluminum conductors are selected and used. Both of these conductors are capable of carrying electric currents with a minimum of heat.

When a conductor is heated, its resistance tends to increase. The result is a higher resistance value. This reduces the amount of current flowing through the conductor to the load. The source tries to force the current through the resistance. The result of this action is a buildup of heat and a power loss at the load. Electric power is figured by the formula $P = E \times I$. It may also be figured by using the formula $P = I^2 R$. This formula is the result of substituting $I \times R$ for E and combining terms:

$$P = E \times I, \quad E = I \times R$$
$$P = (I \times R) \times I$$
$$= I^2 R$$

The application of this formula shows that the power in the circuit is directly related to the square of the current flow through a value of resistance. Power may be in the form of heat. If this is true, the increase of current squared produces quantities of heat.

This may be summarized as follows:

1. The amount of heat produced by resistance heating is directly proportional to the current flow through the resistive material. As the current increases, more heat is created.

2. The current flow through any resistive circuit is proportional to the applied voltage. The applied voltage is a very important factor when determining the amount of heating.

3. Both current and voltage have an influence on the amount of heating. Electric power is the product of voltage and current in the resistive circuit. Power consumption becomes proportional to the amount of heat in the circuit.

Operating Characteristics

One characteristic of heating elements and devices that use any sort of heating element is that the dc resistance of the device when it is cold is very low. The heating element's resistance increases as it heats. If one were to measure the resistance of any heating element when it was cold and then calculate the current flow, the answer would be a very high amount of current. The current for a standard 100-watt lamp would be about 12 amperes. This is much greater than the operating current of the lamp. Operating current for a 100-watt bulb is slightly more than 0.8 ampere. When the light bulb is first turned on, there is a surge of current through it for 1 or 2 seconds. As the filament in the bulb heats, the resistance of the filament also increases. The calculated resistance of a lighted 100-watt bulb is just over 145 ohms. The cold resistance of the bulb is around 10 ohms. This is an operating characteristic of most resistive devices that are designed to use the heat produced by current flow. The same characteristic is true for a radio or television tube filament and the heating element in an appliance. This resulting surge of current is quite typical in resistance heating devices. It is illustrated in Figure 14-1.

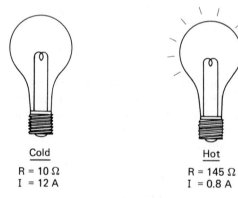

Cold

R = 10 Ω
I = 12 A

Hot

R = 145 Ω
I = 0.8 A

Figure 14-1. A lamp bulb is a resistance load. The amount of resistance varies depending upon the state of the lamp filament.

Resistance Wire

The type of material used as a resistance heating element is important. Copper wire is not used for this purpose. The type of wire used as a resistance heater must have three characteristics: high electrical resistance, high resistance to corrosion, and a high melting point. The wire that meets these criteria will provide a lot of heat with a minimal amount of current flow. The wire that is in use for this purpose is an alloy of nickel and chromium called *nichrome*. Nichrome alloy metals are produced with a variety of shapes. Two common forms are a wire form and a flat ribbonlike form. The resistance of copper wire of about 20 gauge is 0.06 ohm per foot. A comparable nichrome wire has a resistance of 0.66 ohm per foot. Nichrome also has the characteristic of increasing in resistance as it heats.

Nichrome heating wire is used in small appliances in either wire or ribbon form. Larger appliances use a modified form of this wire. This form is housed in a tubular protective cover. The common name for this cover is calrod. An illustration of the cross section of this wire is shown in Figure 14-2. The nichrome element is in the form of

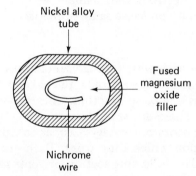

Nickel alloy tube

Fused magnesium oxide filler

Nichrome wire

Figure 14–2. Cross-sectional view of a calrod heating element.

a coil in the middle of the tubular rod. The filler for the tube is a magnesium oxide material. This material is heated. The heated material fuses into a ceramiclike material. This material has excellent heat transfer qualities and also is an excellent nonconductor of electricity. The heat from the nichrome element is conducted to the nickel alloy tubing through this ceramic material. Appliances that use this form of resistive heating element include dishwashers and stoves. The nichrome heating element has basic resistive qualities. In many cases the wire is wound in the form of a coil. The spacing between the turns of wire in the coil are such that there is very little inductance in the coil. The frequency at which the heater functions, 60 hertz, also reduces any inductive effects of the coil.

Lamps

Another device that exhibits resistive qualities is the incandescent lamp. This lamp is made with a clear or frosted glass shell. The filament is made of a resistance metal. This metal has resistive qualities that are similar to nichrome. As a current flows through this wire, it heats. The type of wire selected is designed to glow when it is heated. The metal operates in a vacuum atmosphere. This vacuum will not permit oxidation of the metal due to heat. Oxidation of the metal will make it burn up very quickly. The size of the metal filament determines the amount of light the bulb will produce. The bulb is designed to operate as a resistive load at a power line frequency of 60 hertz. At higher frequencies the bulb will exhibit inductive properties.

The heating element and the incandescent lamp bulb are two basic resistive heating devices. Both have similar characteristics. The heating element uses a nichrome metal. The metal is heated by a current flow. The result is a reddish colored glow on the wire. This material has a relatively long life in open air. The lamp bulb filament requires operation in a vacuum. If it is operated in open air, the metal filament will oxidize and burn up. This destroys the filament wire. There are other resistive loads. Many operate on these same principles.

INDUCTIVE LOADS

Many of the electrical loads with which we are familiar use inductive properties. These include motors, solenoids, relays, and timers. Each of these devices performs one basic function. This is to convert electrical energy into motion. There are several types of electric motors in common use. These motors are categorized as to operating voltage, type, and principle upon which they work. Each group has its own set of characteristics. The following section presents material related to each basic type of motor.

Motor Theory

In general, all motors operate on the same basic theory. This theory is the principle of magnetic attraction and repulsion. Every magnet develops magnetic lines of force. These lines of force move from one pole to the other poles. If one were to place a wire in between the poles, as shown in Figure 14-3, the lines of force would pass across the wire. If the wire was connected to a source of power, a magnetic field would form around the wire. The movement and the direction of the force field in the wire are dependent upon the polarity of the voltage applied to the wire. Parts (b) and (c) illustrate this point.

Each line of force from the magnetic field of the magnet acts as if it was stretched. The interaction of the magnetic field of the wire moves the lines of force away from their original position. The lines of force try to straighten and return to their original position. The result is a force that gives motion to the wire, as shown in Figure 14-4.

Change the single wire that was placed between the magnet's poles

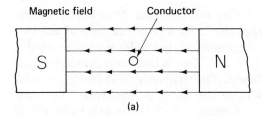

Magnetic field   Conductor

(a)

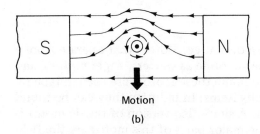

Motion

(b)

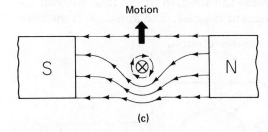

Motion

(c)

Figure 14-3. Interaction between two magnetic fields at right angles to each other.

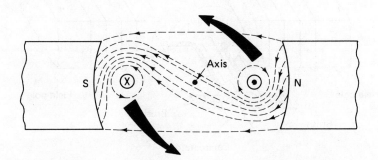

Axis

Figure 14-4. Interaction of a motor armature's magnetic field and the field in the stationary magnet of the motor.

to a loop of wire. This loop is able to rotate on its axis. An electric current is made to flow through the loop. The interaction of the two magnetic fields produces a rotary motion on the loop of wire since it is the only part that is able to move. This is the basic principle upon which the electric motor is based. The magnetic field from the poles influences the magnetic field from the loop, or armature. The armature winding is able to rotate. A shaft is part of the axis upon which it rotates. The shaft is connected to a machine. The rotation of the shaft permits work to be done by the machine. This basic principle is true for all rotating motors.

DC Motors

The dc motor operates from a dc source. Motors that operate from a dc source are found in the automobile as starters, wiper motors, and seat or window motors. In the home these motors are used in tape recorders and in a variety of hobby items. In industry they can be found in diesel electric locomotives. A simplified version of the dc motor is illustrated in Figure 14-5. The major parts of this motor are the field magnet, the armature, a commutator, and a set of brushes. The field is often a combination of a permanent magnet and an electromagnet. This is done in order to increase the strength of the field without increasing the size of the permanent magnet. The armature is the core

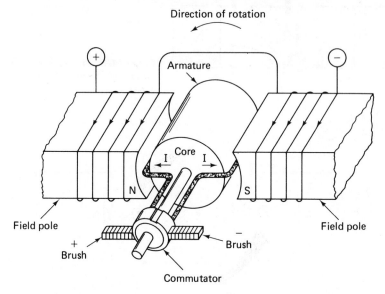

Figure 14-5.   A DC motor, showing all major components.

for the rotating loop of wire. The core provides a stronger magnetic field for the armature without increasing the size of the loop of wire. The commutator is designed and positioned in a manner that maintains the correct direction of current flow from the source to the armature. Connection is made from the source to the commutator by means of carbon brushes.

Motor action occurs as the magnetic field on the armature reacts with the magnetic field in the field windings. The attraction and repulsion of these two magnetic fields forces the armature to rotate. The speed at which it rotates is dependent upon the size of the two magnetic fields. All practical dc motors have several sets of armature windings. This is done to increase the efficiency of the motor.

One factor that is used to advantage in the motor is called counter EMF. This counter EMF develops as the motor armature rotates. The armature acts as a generator. It produces a magnetic field that reacts on the motor field windings. The counter EMF is of an opposite polarity to the applied source voltage. The two voltages add and the net result is a reduced operating voltage and less current flow through the motor. The source voltage is always larger than the counter EMF voltage in a practical circuit. This provides the required current to make the armature rotate. The lack of a counter EMF in a motor results in a very high starting current value for the motor. Once the operational speed is reached, the counter EMF is developed. This establishes a reduced operating current value.

Factors affecting motor action. Certain factors have to be considered for any motor application. These factors are as follows:

1. The speed of the rotation of the motor is inversely proportional to the strength of the magnetic field in which the armature rotates.
2. The torque, or rotational force, of the motor is directly proportional to the current in the armature and the strength of the magnetic field in which the armature rotates.
3. The counter EMF produced is directly proportional to the speed and the magnetic field's strength.
4. As the load on the motor increases, the motor's speed tends to slow down. A decrease in load will tend to increase the motor speed.

Types of dc motors. There are three types of dc motors. The name for each type indicates the relationship of wiring connections between source, armature, and field coils. These are illustrated in Figure 14-6. Each of these motors has its own characteristics, which can be summarized as follows:

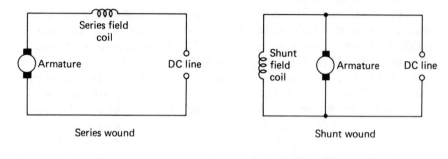

Series wound   Shunt wound

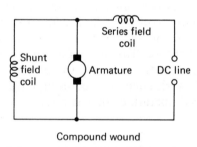

Compound wound

Figure 14–6. Three wiring configurations for DC motors. Each is discussed in the text.

1. **Series motor.** High starting current, high starting torque, increased load slows motor, decreased load speeds motor, no load will make motor "run away" and possibly fly apart physically. This motor must be directly connected to the load.

2. **Shunt motor.** Low starting current, low starting torque, fairly constant speed under varying load conditions, safe no-load speed; will not "run away."

3. **Compound motor.** The best of both motors. Fairly high starting current and torque, good speed regulation under a variety of load conditions, and will not "run away" under a no-load condition.

The selection of a specific dc motor is dependent upon the application for which it is intended. There are situations that require a series motor and a shunt motor. When this is the case, a set of contact assemblies is used to switch from one wiring connection to the other. This is done because the compound-wired motor is not capable of performing the required amount of work.

AC Motors

The ac motor has gained wide acceptance because of the availability of ac power. Almost all home and industrial motors are operated from ac sources. These motors are also categorized. There are two general groups for ac motors: single-phase motors and polyphase motors. The single-phase motors are used in applications that require less than 1 horsepower of energy. (One horsepower is equal to 746 watts of electric energy.) These motors operate smaller devices. Larger ac motors that operate from polyphase sources are subdivided into either induction or synchronous groupings.

Polyphase motors. One of the most commonly used ac motors is the induction motor. A sketch of the motor is shown in Figure 14-7. The induction motor is very similar to a transformer in the way it works. One winding is wound on two opposing poles of an iron ring. This winding is called the *stator*. It is the stationary winding of the motor. This winding is connected to the ac source. It acts in a manner similar to the primary winding of a transformer.

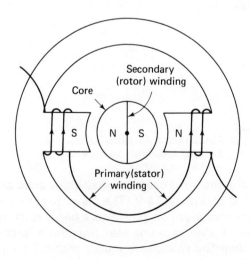

Figure 14-7. Wiring of an AC induction motor.

The rotating portion of the induction motor is called the *rotor*. It is made of a loop of wire that is wound on an iron core. When the source voltage is connected to the stator winding, magnetic poles are established. These can be seen in Figure 14-7. One pole becomes the north pole and the other pole becomes the south pole of the stator. There is a magnetic field produced between these two poles. The magnetic field crosses the closed loop of the rotor. This induces a magnetic field in the rotor. The rotor's magnetic field is of opposite polarity to the stator's magnetic field. This results in the attraction of the north pole of the stator and the south pole of the rotor. The opposite set of poles are also attracted to each other.

As an ac current flows through the stator winding, its poles will alternate in polarity. The end result is a rotor that is locked in place with the stator. The motor does not rotate under these conditions. The solution to this problem is to make one of the magnetic fields rotate. If this can be accomplished, the rotor will rotate in step with the magnetic field.

In a practical induction motor of this type there are two sets of poles and windings. These sets are 90 degrees apart on the stator. Each set is connected to the same power source. One set of windings is called the *starting* winding. The second set of windings is called the *running* winding. The starting winding is used to establish a second magnetic field. This field starts the rotor in motion. A schematic of this type of motor is shown in Figure 14-8. Once the rotor starts to turn, its rotational force tends to "chase" the magnetic field of the stator and it keeps on rotating.

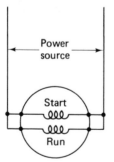

Figure 14-8. An induction motor using a rotating magnetic field as a starting winding.

A second type of induction motor is called the capacitor-start induction motor. It is illustrated in Figure 14-9. The capacitor is connected in series with the start winding. This produces a phase shift in the start winding. This type of start winding also provides a larger amount of starting torque than the split-phase motor without a capacitor.

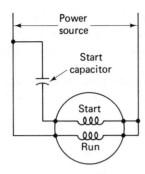

Figure 14-9. An induction using a capacitor that provides a phase shift for starting rotation.

Another type of induction motor is called a shaded-pole motor. This motor works on the same principle as does the split-phase motor. The difference between these two types of motors is that the shaded-pole motor has one or two copper bands wrapped around its rotor. The copper bands form a conductive secondary path in the rotor. The induced current in these copper bands produces the phase shift required to start rotation.

The shaded-pole motor has less starting torque than other induction motors. These motors tend to be self-regulating as to their speed. Most of these motors handle light mechanical loads. They are used where precise speed is required. One such application is the motors used in timing devices.

One problem that arises when using a split-phase motor is what to do with the starting winding once the motor is running at speed. This winding uses electric power. This power is wasted once speed is attained. Most induction motors have a method of disconnecting the starting winding once the motor is running. In fact, there are two common methods in use. One of these uses a centrifugal switch. This is illustrated in Figure 14-10. Under start conditions the centrifugal switch contacts are closed. This provides a current path to the starting winding. As the rotor moves up to full speed, the centrifugal

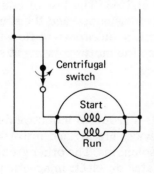

Figure 14-10. A centrifugal switch disconnects the starting winding as the motor gets up to full speed.

forces acting on this switch separate the contacts. This occurs at about 75% of full speed. Since this switch acts in relation to motor speed, it is designed to close when rotor speed drops to under 30% of running speed.

Another system used to control current flow through the starting winding is shown in Figure 14-11. This system uses a device called

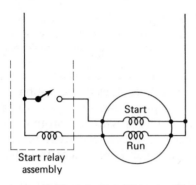

Figure 14-11. A relay often replaces the centrifugal switch starter.

a relay to switch this circuit on and off. The relay has two parts. One is a coil that forms an electromagnet. The second part of the relay is a set of contacts. The contacts touch when the coil is energized. The coil of the relay is wired in series with the running winding of the motor. The contacts of the relay are wired in series with the start winding. These contacts are normally in the off position.

When a source voltage is first connected to the motor, a large amount of current flows through the motor. This current energizes the relay coil. The magnetic field in the relay coil closes the contacts and completes the start circuit. The current flows through the start winding and the motor starts its rotation. The current flow in the stator is reduced as the motor builds up to its operational speed. The counter EMF produced by the rotation of the rotor produces an opposing voltage across the stator. The result is a drop in voltage and current, and the relay coil is deenergized. The loss of energy in the relay coil causes the relay contacts to separate, and the start winding is disconnected from the circuit. This reduces current flow in the motor to that which is required to keep the motor rotating at speed.

OTHER INDUCTIVE LOADS

In the electrical field one finds a group of inductive devices that are often used as a load. These devices are divided into two groups. One group of the devices is called solenoids. The other group is called relays. Each of these devices operates on electromagnetic principles.

Each has its own set of characteristics. This section of the chapter covers both of these devices and also timing devices used in the electrical field.

Solenoids

The solenoid uses electromagnetic principles to convert electrical energy into motion. The basic solenoid is shown in Figure 14-12. It consists of an iron frame upon which is wound a coil of wire. The iron frame has a hollow center core. A soft iron rod that fits in the center of the core is placed so that it is almost completely out of the hollow center. When an electric current flows through the coil of wire, a magnetic field is established in the frame. The magnetic field flows better through the soft iron core material than it does through free air. The magnetic field pulls the soft iron core into the frame. It is moved until the soft iron core is completely inside of the magnetic field of the frame. The core is held in this position until the current that formed the field in the frame is stopped.

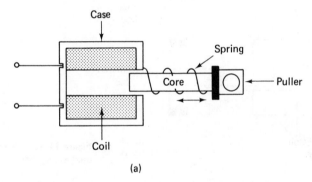

(a)

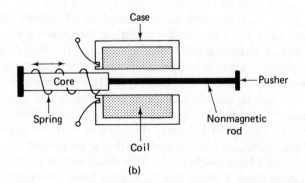

(b)

Figure 14-12. Solenoid action in order to perform work.

When the magnetic field is stopped, the core is free to move. A spring, as shown in Figure 14-12, is attached to the core. This spring is designed to pull the core out of the frame. It returns the core to its original position. The arrangement of the core and spring will always pull the core into the frame when the coil is energized. A device is attached to the core that permits the movement of the core to do some mechanical work.

In many instances the core pulls a mechanism. In other instances the core pushes the mechanism away from its body. In still other situations the core has a set of electrical contact points attached to it. A second set of contact points is mounted on the frame of the solenoid, as shown in Figure 14-13. When the solenoid is turned on, the core action connects the sets of contact points. When the coil is de-energized, the spring separates the contact points. In this type of action the solenoid is used as a switch. The movement of the core activates the switch contacts.

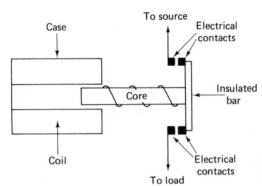

Figure 14-13.   Solenoids are also used as switches.

A typical application for the solenoid switch is found in the automobile. The solenoid coil uses relatively little current. The contacts are designed to carry much heavier amounts of current. The lights of the automobile require much current. Information given previously showed that a large-diameter wire is required to carry large currents. It is too costly to run a large wire from the headlamps to the headlamp switch and then to the battery of the automobile. The switch required to handle the large currents is also large and costly. A better arrangement is shown in Figure 14-14. A solenoid contactor device is placed under the hood between the battery and the headlamps. The switch that energizes the solenoid coil is a small unit. It is placed in the dashboard of the automobile. This small switch controls the headlamp current from a point that is remote from the lamps and source.

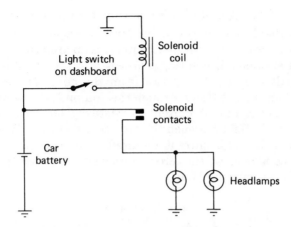

Figure 14-14. Wiring of the starter solenoid circuit of an automobile.

This is one form of remote control. In this situation the primary circuit is the solenoid coil circuit. The secondary circuit has a short run of heavy wire. It is the headlamp circuit. The primary circuit controls the secondary circuit.

A modification of this circuit is found in motor contactor assemblies. A solenoid coil assembly has two or more sets of contact points attached to its plunger. Each set of contact points is connected to one wire from the source. The other side of each set is connected to the motor windings. When the "on" switch is pushed, the coil is energized. This connects the sets of contact points and turns on the motor. If power to the contactor assembly is interrupted, the spring attached to the contact points pulls the points apart. The spring holds the points apart and keeps the circuit off until the "on" button is pushed again. This is a fail-safe arrangement that disconnects power to the machine until the operator turns it on again. If this was not used, the operator could think that the machine was jammed. The operator could start to investigate to see if repair could be made. A return of power to the machine circuit would make the machine operative and the operator would be injured.

Relays

A relay operates in a manner that is very similar to the solenoid. The major difference between the two devices is the manner in which the contact points are moved. Both the solenoid and the relay operate on the principles of electromagnetism. An illustration of a relay is shown in Figure 14-15. A coil of wire is wound on a permanent core. The core and the frame of the relay form a magnetic circuit. Also included as a part of the magnetic circuit is an armature. The armature is pivoted at one end. A spring is attached to the armature so that the spring holds the armature away from the core. The magnetic field created when current flows through the coil pulls the armature against the core. Initial circuit current is relatively large to overcome the weak magnetic field between the armature and core. When these two parts touch, the coil current required to hold them together is reduced. This is because the direct connection between the metal parts makes a better magnetic circuit.

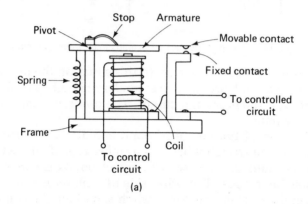

(a)

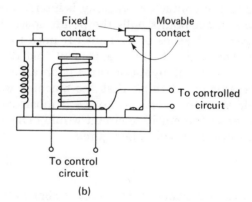

(b)

Figure 14-15. Construction of a relay. Both normally open contacts (a) and normally closed contacts (b) are shown.

A contact point is attached to the armature of the relay. Its mate is attached to the frame of the relay. These contact points are often electrically insulated from the member that holds them. The contact points are designed to make (connect) or break (separate) as required by the circuit. The relay coil is wired into a control circuit. A switch or other sensing device is used to complete the control circuit wiring to a power source. This is illustrated in Figure 14-16.

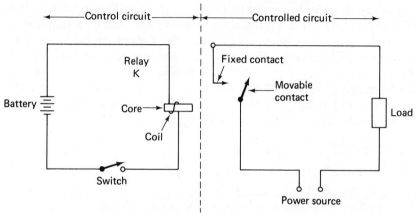

Figure 14-16. Relay contact circuits are used to operate secondary circuits.

In the case of heavy current load requirements and/or higher voltage requirements, the contact points complete a secondary circuit. The switch and coil circuit control the operation of the secondary circuit.

Contact point sets. Relays and solenoids can be manufactured with a variety of contact point arrangements. The simplest of these is called a normally open (N.O.) arrangement. Various contact point arrangements are shown in Figure 14-17. A second basic arrangement is described as normally closed (N.C.). The longer line on the drawing represents the movable armature in these drawings. The shorter arm is used to show a fixed contact point. When the arrow touches the

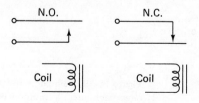

Figure 14-17. Contact arrangements for relays. Both normally open and normally closed contacts are illustrated.

longer arm, the indication is that a connection is made. The contact points are always shown in the "off" position of the relay or solenoid. Energizing the relay moves the armature toward the symbol for its coil.

Each set of contact points is called a *pole*. Thus, a single-pole relay has one set of contact points. The contacts are also described by the number of positions, or *throws*, of the unit. One may have a single-pole double-throw contact set, as shown in Figure 14-18. More than one set of contact points are indicated by use of a number. A four-pole switch has four sets of contacts. Relays are built with one or more sets of contacts. The exact number depends upon the switching requirements of the circuit and the capabilities of the mechanism of the relay to move all the contact sets. The dashed line on the drawing connecting the two armature lines shows that these are mechanically connected but not electrically connected.

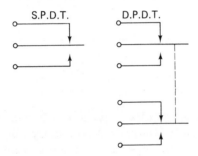

Figure 14-18. Multiple contact arrangements for relays.

Timers

Many electrical devices require some form of timing circuitry. Electric timers are often driven by a very slow speed motor. In some cases. the speed reduction assembly is connected between the motor and the timing mechanism. Timers are basically found in one of two forms. The first is shown in Figure 14-19. An offset cam rotates on the motor drive shaft. The cam is designed so that it will close a set of contact points during its cycle. The contact points are wired to

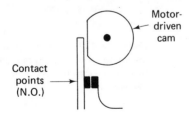

Figure 14-19. Simple timer arrangement uses a motor-driven cam to operate circuit contacts.

the circuit, which has a timed operational cycle. The speed of the motor and the shape of the cam determine how often the contact points are closed and the circuit is operational.

A second form of timer uses a wheel to which copper foils are attached. The foils have contact arms riding against them. The foil travels in a circular direction. Portions of the circle have foil and portions are blank. When the contact arm touches the foil, the circuit is completed. When the arm is in contact with the blank section, there is no electrical contact. This principle is shown in Figure 14-20. The number of contacts is dependent upon the size of the disc and the ability of the motor to make it turn. A device that uses this principle is the automatic washing machine timer assembly. In some cases the timer is linear (straight) rather than circular in form. The mechanism works in the same manner in either case.

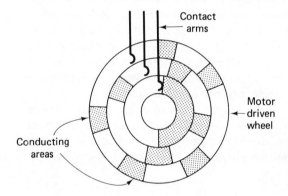

Figure 14-20. More complex timers have sets of contacts that ride on a wheel. The circuit is completed when the contact touches the conductive area.

SUMMARY

Almost all the machines used in the home or in industry require a source of electric power. Each consumer of electrical energy is classified as a load. Each load requires a specific amount of electric current at some operating voltage. The power formula $P = E \times I$ is used when calculating electric power consumption. Current flowing through a conductor produces heat. This heat may be an undesired feature. It wastes power that could be used in the performance of work. There are times, however, when heat is the desired outcome of the circuit. One device that converts electrical energy into heat energy is the heating coil. This coil is made from a resistance wire called nichrome. Nichrome wire is designed to produce heat when an electrical current flows through it. The wire is considered to be resistive in nature. A second device that is also resistive in nature is the light bulb. The light bulb has a filament that glows to a bright light when it is heated. Both of these types of resistive loads have a low resistance when cold and a higher resistance when heated.

A second type of load classification is the inductive load. These loads form electromagnetic fields as they operate. Inductive loads include motors, solenoids, relays, and timers. Each converts electrical energy into rotary motion. Motors may operate from ac or dc sources. The ac motor requires a starting winding. The starting winding changes the phase relationship between the stator and the rotor of the motor. Starting windings are disconnected once the motor is running at its operational speed.

Two other devices that use electromagnetic principles are the solenoid and the relay. Both have a coil that is energized by a control circuit. The resulting electromagnetic field moves a core or an armature. Solenoid action is often used to move mechanical parts. Solenoids may be used as switches if contact points are included in the assembly. Relays operate sets of contact points. The contact points control a secondary circuit. This circuit is not electrically connected to the control circuit in most applications. The result is a primary control circuit that operates a secondary load circuit.

A timer is a motor-driven switch. The rotation of the motor drives either a wheel or a cam. Contact point assemblies are attached to these rotating parts. The motor normally operates at a very slow speed. As the contacts open or close, the load circuit is controlled.

QUESTIONS

14-1.   A resistive load is

     a. a lamp

     b. a heating element

     c. a motor

     d. all of the above

     e. answers a and b above

14-2.   Motor theory is based upon

     a. magnetic repulsion

     b. magnetic attraction

     c. rotation of the earth's magnetic field

     d. answers a and b above

14-3. Counter EMF is produced by

    a. fixed magnetic fields

    b. armature reaction

    c. rotating magnetic fields

    d. field coil magnetism

14-4. A starting winding is used to

    a. stop rotational motion

    b. as a brake for a motor

    c. increase current flow

    d. start rotational motion

14-5. Start windings are disconnected by

    a. relays

    b. centrifugal contact assemblies

    c. electromagnetic action

    d. all of the above

14-6. In the solenoid, the

    a. frame forms a magnetic field

    b. core of the magnet moves

    c. armature of the magnet moves

    d. moving part performs no work

14-7. Solenoids are used for

    a. pushing action

    b. pulling action

    c. switch contact action

    d. all of the above

    e. answers a and b above

14-8. In the relay, the

    a. frame forms a magnetic field

    b. core of the magnet moves

    c. armature of the magnet moves

    d. moving part performs no work

14-9. Contact points that are touching in the "at rest" position are called

    a. normally closed

    b. normally open

     c. circuit makers

     d. circuit breakers

14-10. The sets of contact assemblies are called

     a. contactor sets

     b. poles

     c. magnetic connectors

     d. relay points

15

Test and Measuring Instruments

Each person involved with electrical work will undoubtedly have to measure some quantity of electricity. There are several different types of measuring devices in common use today. The measuring industry is going through a change of state at the present time. For years analog types of measuring devices were used. These devices converted the electrical quantity being measured into a reading found on a precalibrated scale of a meter. These types of measuring devices are still in use. They are gradually being replaced by digital readout types of measuring devices. Both analog and digital style meters are shown in Figure 15-1.

BASIC METER MOVEMENT

One of the most popular meter movements in use today is the d'Arsonval type of movement. This meter movement is similar in operational theory to that of the electric motor. A sketch of this type of meter movement is shown in Figure 15-2. The movement consists of a permanent magnet, an armature, and a pointer. A set of coil springs is attached to the armature. The springs hold the armature in the rest position when the meter is not connected to a circuit.

When the armature is connected to a test circuit, an electric current flows through its winding. This forms an electromagnet. The magnetic field of the armature reacts with the magnetic field established by the permanent magnet's pole pieces. The armature's field causes it to rotate. The rotation of the armature swings the pointer

Figure 15-1. Meter faces in current use include analog styles with a needle and scale (a) and digital readout types (b). (Courtesy Simpson Electric Co.)

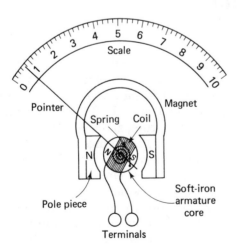

Figure 15-2. Construction of an analog meter movement.

needle to some position other than the "rest" position on the left side of the scale. The amount of current flowing through the armature will move the pointer to any point on the scale. As more current flows, the pointer swings closer to the right side of the scale. At maximum current flow, the pointer is on the right side of the scale. The amount of current flowing through the armature coil in this type of meter determines the amount of deflection. Armatures are produced with various current ratings.

This type of meter movement has some limitations. These include the overall current limitations based on the size of the wire in the armature, the deflection of the needle is limited to only one direction, this meter movement will only register a direct current, and the voltage required for deflection is very small. There are many methods in use today that modify the above limitations. These methods extend the capability of the meter movement. Let's review each of these and see how the modifications apply.

Current Limitations

Armature wire size will limit current flow in the meter. A meter movement's range is extended when a portion of the current is passed around the meter armature. This is illustrated in Figure 15-3. One or more resistance values are wired so that they are in parallel with the meter movement. If the value of R_1 is equal to the ohmic value of the armature, the current will divide equally between the two resistances. If the value of R_2 is smaller than the ohmic value of the meter movement, a greater percentage of the total circuit current will flow through R_2 than flows through the meter. R_3, being even a smaller value than R_1 or R_2, will allow even more current to flow in the circuit. These resistors are called *shunt* resistors. They permit a calculated percentage of current to bypass, or shunt, the meter movement. In this way the current capacity of the meter is extended.

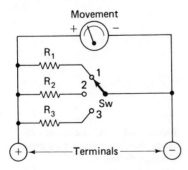

Figure 15-3. Current ranges are increased by use of a shunt to bypass part of the current.

The scale on the face of the meter would have to be recalibrated to show these increased values.

Directional Limitations

The mechanical limitations of the springs, pointer dial, and armature will permit meter deflection in only one direction. This is from left to right. If a connection to a circuit was such that the pointer attempted to move in a counterclockwise direction, the pointer could be permanently damaged. Correct polarity between the circuit under test and the meter movement must be observed. Incorrect connections require that the leads to the meter be reversed in order to have proper deflection. Some meters have a reversing switch that changes lead polarity without moving the leads.

Direct Current Limitations

The meter is designed to react to a current that flows only in one direction. An alternating current will make the armature oscillate at the frequency of the ac. This limitation is overcome by the addition to the meter circuit of a device called a *rectifier*. The rectifier changes the ac waves into dc waves. These dc waves flow only in one direction. They can be measured by the meter movement.

Voltage Limitations

Assume that a model meter has an armature resistance of 100 ohms. The current for full-scale deflection is 0.001 ampere. Use of Ohm's law shows that the voltage required to move this current through 100 ohms is 0.1 volt. This meter may be used as a voltage-measuring meter if the voltage of the circuit under test is dropped to all but 0.1 volt. A circuit showing how this is accomplished is shown in Figure 15-4. A resistance value is connected in series with the meter

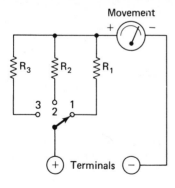

Figure 15-4. A multiplier extends the voltage-measuring capability of the meter.

movement. The resistance of the series resistor is much higher than the 100 ohms of the meter. Most of the voltage in the circuit drops across the larger resistor. The balance drops across the resistance of the meter movement. This makes a current flow through the meter. Application of Ohm's law for the current and resistance in the armature of the meter will give a voltage value. The meter scale may be calibrated in units of the volt when it is used in this manner. Different values of resistance as shown by R_2 and R_3 in the circuit will change the range of the meter to other values of voltage.

Other Limitations

The basic meter movement can be modified to measure other quantities. One of these is resistance. Two additional parts are required. One is an internal power source and the other is a calibrating resistor. The circuit for a resistance measuring meter, of ohmmeter, is shown in Figure 15-5. If a known voltage is applied to a specific resistance, the unit of resistance may be determined by use of Ohm's law. This circuit uses this concept in order to measure amounts of resistance. The circuit is completed by connecting a wire between the test terminals. A full-scale current reading is obtained by adjusting the calibration control. An unknown value of resistance is then connected between the test terminals. A lesser amount of current will now flow in the circuit because of the additional resistance in this circuit. The scale is calibrated in units of the ohm to show the amount of resistance being measured.

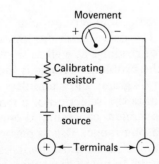

Figure 15-5. An ohmmeter requires its own power source.

Another measurement that may be made with the meter is temperature. In this type of measurement a device called a *thermocouple* changes the heat energy into electrical energy. The thermocouple is constructed of two dissimilar pieces of metal wire. These wires are twisted together at one end. This is the end that is placed in the heated area. The other ends of the wires are connected to a meter.

This is illustrated in Figure 15-6. Heating the ends of the wires produces a voltage difference on the wires. This is measurable. The meter reads the current flow produced by the difference of potential between the two wires. The scale of the meter is calibrated in units of temperature rather than voltage or current.

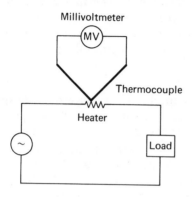

Figure 15-6. The thermocouple meter measures the voltage produced by heating the junction of a bimetal strip.

In every situation described the meter is really measuring the flow of current in the circuit. The scale on the face of the meter may show other quantities, but in reality a current flowing through the armature of the meter movement develops a magnetic field. This magnetic field reacts with the permanent field of the magnet. The result is a rotary motion similar to that of the induction motor.

Multimeters

A variation of the basic meter that has become very popular over the past several years is the multimeter. The basic meter has limitations. These are related to a specific range of readings as well as what is being measured. A partial solution to the problem of range is shown in Figures 15-3 and 15-4. Both of these diagrams show a multiposition switch in the circuit. The switch permits the selection of a range-extending resistor. The basic meter movement capability is enhanced when it is able to measure more than one range. This is particularly true if the meter is a portable type. Some meter movements are permanently installed in an operational unit. In many cases these meters only measure one quantity and have only one range of readings. Many technicians have a need to measure several values of a current or voltage. When this is required, a meter with range-switching capabilities is preferred.

The meter illustrated in Figure 15-7 is called a multimeter. This meter is wired with a multiple range selector switch. It is capable of measuring ac voltage, dc voltage, dc current, and dc resistance. Selec-

Figure 15-7. A typical multimeter as used in industry. (Courtesy Simpson Electric Co.)

tion of the proper function switch permits this versatility. The voltage ranges cover low values of 1 or 2 volts up to a high value of 1,000 volts. This type of meter is desired as a portable test meter because of its versatility. One unit has the capability of making a wide variety of measurements.

This type of meter has a very significant limitation. This relates to the effect of the meter on the circuit being measured. The effect is called the *loading* effect of the meter. Each measuring device has some resistance. A voltmeter is rated at a quantity of "ohms per volt." Inexpensive voltmeters are rated at 1,000 ohms per volt. This means that a meter range of 20 volts will have a resistance of 20,000 volts ($20 \times 1,000 = 20,000$). The diagram in Figure 15-8 illustrates a circuit with two 10-kilohm resistances in series across a 40-volt source. Operating characteristics for this circuit are determined by use of Ohm's law. Total current flow is 0.001 ampere. The voltage drop across each resistor is 20 volts. A meter that has a resistance of 20 kilohms is used to measure the voltage drop developed across one of

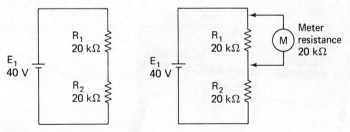

Figure 15-8. The resistance of the meter may affect the readings and operation of the circuit.

the resistors. The resistance value across this resistor changes owing to the addition of the parallel resistance of the meter. The total resistance across this resistor is now 10 kilohms. Total circuit increases from 0.001 to 0.0013 ampere. The voltage drop across the resistor has decreased from 20 volts to a new value of 13.3 volts. If the person making the voltage reading is not aware of this loading effect of the meter, it would be easy to assume that the circuit was malfunctioning.

Better quality multimeters have a much higher ohms per volt rating. Most of these meters use a value of either 20 or 30 kilohms per volt. This value is called the *input sensitivity* of the meter.

ELECTRONIC METERS

Meter input sensitivities that are constantly high are desired for work on low-voltage and low-current circuits. Meters with an input sensitivity of 1 or 2 megohms for each range are used for measuring a wide variety of circuit values. These meters are classified as electronic multimeters. Older versions of these meters used vacuum tubes as amplifiers to raise the input resistance value. These are called VTVM's. This stands for vacuum-tube voltmeter. Many of these types of meters in current production use solid-state devices such as transistors or integrated circuits instead of vacuum tubes for the amplifiers. These meters are called EVM's or TRVM's. These letters indicate an electronic or a transistorized voltmeter. A picture of this type of meter is shown in Figure 15-9.

Some electronic meters do not measure current. This is a design characteristic. Many of the newer EVM's have a digital readout instead of the more traditional analog meter movement. The digital type of readout offers a more precise reading than can be obtained from an analog dial. Digital readout meters, having more complex circuitry, are normally more expensive than their analog counterparts.

Figure 15-9. An electronic meter that utilizes a digital readout. (Courtesy Simpson Electric Co.)

There are times when an analog type of meter movement is mandated. In some circuitry the technician has to adjust components for either a maximum or a minimum value. The fairly slow response of the digital meter makes it difficult to use in these cases. The analog type of meter movement displays changes more conveniently. Some digital meters have an accessory analog display just for this purpose.

CONNECTING METERS INTO THE CIRCUIT

The only way to obtain correct readings of circuit values is to properly install the meter in the circuit. This is true for both permanent installations and test units. An incorrect connection may often damage the meter. Meters are expensive and not always readily available. Doing without a meter while it is out of service and being repaired is a great inconvenience to the technician. It is much better to be certain of how to connect the meter into the circuit than it is to be sorry after the meter is damaged.

Voltmeters

Probably the most common measured quantity is voltage. Voltage by definition is a difference in electrical pressure between two points in a circuit. Voltage, therefore, is measured between two points in a circuit. Voltmeters have two leads that connect the meter to the circuit. One of these leads is marked with a minus (-) sign. It usually is the black-colored lead of the meter. The other lead is marked with a plus (+) sign. It usually is the red-colored lead from the meter. Proper polarity must be observed when making voltage readings.

When making a voltage reading with a multimeter, the following steps must be observed:

1. Selection of a range on the meter that will safely measure the voltage.
2. Selection of the proper function. Multimeters will measure both ac and dc voltage values.
3. Correct connection of the meter leads to the circuit. The positive lead must be connected to the most positive point to be measured. This is illustrated in Figure 15-10. Each component has a definite polarity in a dc circuit. This refers to the laws established by Kirchhoff. The negative meter lead is connected to a less positive (or more negative) point in the circuit. Correct polarity connections give the meter the proper deflection movement and do not damage the meter. In other words, the voltmeter is always

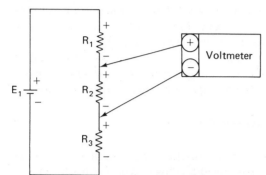

Figure 15–10. The voltage-reading meter's leads are placed across the component when a voltage measurement is made.

connected in parallel with the component whose voltage drop is to be measured. (This step is only true when measuring dc voltage values. If one is measuring ac voltages, there is no need to observe correct polarity. This is because the polarity of the ac voltage is constantly changing.)

4. It is wise to consider the value of making connections between the circuit and the meter when the operating power is turned off. This is particularly true when working with voltages that are above 100 volts. Turn off the circuit power, make the meter connections, and then turn the power on again. The meter will indicate the measured voltage. This method may be slower than working under "power on" conditions, but it is definitely safer.

Ammeters

The measurement of circuit current requires an approach that differs from the methods used to measure voltage. Current is the flow, or movement, of electrons in the circuit past a given point during a time period of 1 second. Current measurements are made in one of three ways. One method uses a multimeter or ammeter. This method requires the installation of the meter as a part of the circuit. The second method uses the principles of electromagnetic induction in order to make a measurement. The third method relies on the use of Ohm's law.

The first requires the disconnecting of the circuit at some point. Most circuits in which current is to be measured are wired as a series circuit. One such circuit is shown in Figure 15-11. An ammeter or multimeter's leads have to be connected in such a way that all the circuit current flows through the meter. Current in the series circuit is the same in all parts of the circuit. This means that the circuit may be broken at any point and the ammeter added to the circuit. The following steps are required:

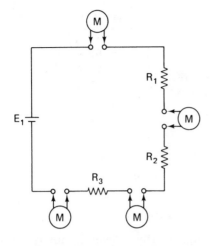

Figure 15-11. Current-reading meters (ammeters) must be wired into the circuit to measure current flow.

1. Turn off circuit power.
2. Select a convenient point in the circuit to make the measurement. One suggestion when working with high-voltage circuits is to try to find a point close to circuit common, or circuit negative, for meter connection. The voltages are lower at these points. This keeps the higher-voltage values away from you and the meter. This is a safety suggestion.
3. Select the proper range for the meter reading. The electrical specifications for the circuit should provide some idea of normal voltage and current values.
4. Select the proper function. Many multimeters do not measure an alternating current. When doing this, be certain that the meter has this capability.
5. Disconnect the circuit at the point you wish to insert the meter. Add the meter leads at the break. This is shown in Figure 15-11. Be sure to observe correct polarity. The positive lead from the meter is connected to the most positive side of the circuit.
6. Turn on circuit power. Measure the value of current. A word of caution at this point is in order. Be prepared to turn off the circuit power if there is a greater amount of current flow than expected. A short-circuited component in the circuit reduces total circuit resistance. The result is a greater amount of current than expected in the circuit. See Figure 15-12 for an illustration of this. The action of the short-circuited component increases the current in the circuit from an expected value of 0.2 ampere to a new value of 1.0 ampere. This unexpected additional current

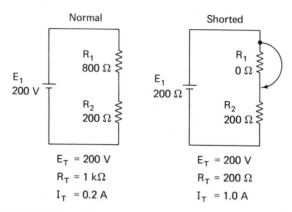

Normal   Shorted

E_1 200 V R_1 800 Ω E_1 200 Ω R_1 0 Ω

              R_2 200 Ω R_2 200 Ω

E_T = 200 V E_T = 200 V
R_T = 1 kΩ R_T = 200 Ω
I_T = 0.2 A I_T = 1.0 A

Figure 15-12.   Effects of a short circuit on measurements and circuit values.

could cause meter damage if the power is not quickly turned off.

The second method of measuring a circuit current is much simpler to use than the previous method. This procedure uses a special type of ammeter. The ammeter, which is shown in Figure 15-13, is actually

Figure 15-13. The clamp-on ammeter is a useful tool for measuring current flow. It works on the principle of transformer action. (Courtesy Amprobe Instrument Co.)

a transformer. The primary of the transformer is the wire in which current is flowing. The circuit current establishes a magnetic field around the wire. The size and the strength of the field are determined by the amount of current in the wire. The jaws of this ammeter will open. This allows the iron core to encircle the wire. The secondary of the ammeter is connected to a meter movement. The voltage induced by transformer action causes a current to flow through the meter movement. The scale of the meter is calibrated to give a correct current value. This is shown in Figure 15-14.

Figure 15-14. An application of the clamp-on ammeter. (Courtesy Amprobe Instrument Co.)

This type of clamp-on meter is very convenient to use. One must be careful to place the clamps around only one wire at a time. Additional wires may have current flows in different directions. This could produce a false reading on the test meter.

A third method of measuring circuit current is also accomplished without disconnecting wires in the circuit. This method requires the use of Ohm's law and is illustrated in Figure 15-15. A voltage drop is measured across a known value of resistance in the circuit. The measured voltage and the known current are used in Ohm's law to find the current. In this circuit, 8 volts measured across a 1-kilohm resistor will produce a current of 0.008 ampere.

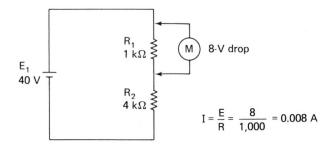

Figure 15-15.   Use of Ohm's law to measure current in a circuit.

A second version of this method is also offered. Assume that the circuit to be measured has a resistance value of 100 ohms for one component in the circuit. Assume, also, that the multimeter being used has a group of voltage ranges that are multiples of 10. This, by the way, is very common. Ohm's law is applied before the circuit values are measured to convert the voltage ranges into current ranges:

$$I = \frac{E}{R}$$

0 – 1 V	=	0 – 0.01 A
0 – 10 V	=	0 – 0.1 A
0 – 100 V	=	0 – 1 A
0 – 1,000 V	=	0 – 10 A

This method is not quite as fast as the use of a clamp-on type of ammeter. It has the advantage of using the multimeter without breaking into the circuit in order to make meter connections. It also allows the measurement of an ac current with a multimeter that is not capable of normally measuring this ac value.

Resistance Meters

The measurement of resistance requires that the circuit power be turned off. It also requires that the component whose resistance value is being measured be disconnected from the circuit. The reason for the first of these conditions is that the ohmmeter has its own power source. The current flow in the ohmmeter circuit is the result of the applied voltage of the meter. This is calculated and converted into a resistance reading. The ohm scale on the meter face shows the result of the calculations. If one were to attempt to measure resistance and had an additional power source in the circuit then all the readings would be incorrect.

The second condition is necessary because of the possibility of obtaining a false reading if the component is left in the circuit. This is illustrated in Figure 15-16. The ohmmeter leads are placed across R_1 to measure its value. A current from the ohmmeter source produces a current flow through this resistance. Current is also flowing through the additional resistance of R_2. This unwanted current flow due to the additional resistor allows a greater current flow than if only R_1 was in the circuit. The additional current flow produces a false reading on the ohmmeter.

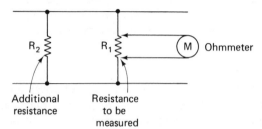

Additional resistance

Resistance to be measured

Figure 15-16. False readings are obtained when attempting to measure resistance without removing the component from the circuit.

The procedure for measurement of circuit resistance is as follows:

1. Disconnect circuit power.
2. Disconnect component from the circuit. Only one end of the component has to be disconnected to make a resistance measurement.
3. Select proper resistance range. Most multimeters show a basic range for resistance. The range switch is used to add "multiplying" ranges. These would be $R \times 10$, $R \times 100$, etc., on a typical meter.
4. Select proper function. This is the resistance or ohms position.
5. Follow specific instructions of the meter manufacturer in order to set the meter for an accurate reading.
6. Reconnect the circuit component.
7. Turn on circuit power as required.

There are instances when one wishes to check for a continuous circuit. This is called a *continuity check*. The ohmmeter may be used to do this. The check for continuity is actually a resistance measurement. The amount of resistance measured in the circuit will give an excellent indication of the quality of the circuit. If the meter shows a very low resistance, as illustrated in Figure 15-17, then the circuit has continuity. If, on the other hand, the circuit has a break in it, the resistance value is very high. Wires and switches are tested in this manner. If the test indicated that there was some resistance but the value was much higher than expected, some component or wire would probably be failing.

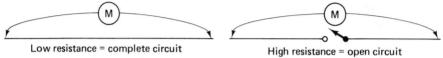

Low resistance = complete circuit   High resistance = open circuit

Figure 15-17. Continuity measurements are determined as shown.

VISUAL READOUT DEVICES

The meters described in the previous sections of this chapter provide a reading of a specifc value. These devices are limited because they can display the quantity only at the moment it occurs. These devices also are limited because they cannot display rapid changes or the rate at which these changes occur. Two families of instruments have these capabilities. These are graphic recorders and the oscilloscopes. Both of these groups have some similar characteristics. Let us look at the graphic recorder first.

Graphic Recorders

Graphic recorders have the capability of displaying values on two axes. One axis is called the vertical axis. The other axis is the horizontal axis. The vertical axis is used to display or record changes in the amplitude of the measured value. The horizontal axis relates these changes to units of time. Figure 15-18 illustrates these axes.

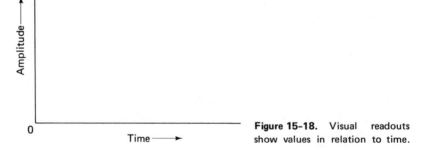

Figure 15-18. Visual readouts show values in relation to time.

The graphic recorder uses a paper chart. A marking device draws a line or lines on the paper chart. These lines relate to the changing values being measured. The paper chart is moved past a given point on the recorder at a constant rate of speed. This speed will be dependent upon the time requirements of the measurement. Speed ranges from 20 to over 3,600 millimeters per hour are common. One type of graphic recorder is shown in Figure 15-19.

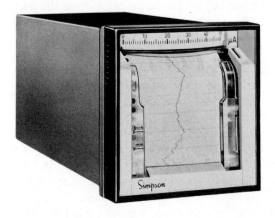

Figure 15-19. A chart recorder records circuit values on a paper chart. (Courtesy Simpson Electric Co.)

Typical connections to this recorder are made to the vertical indicator section. Variations in the amplitude of the measured quantity cause the marking device on the chart to draw a corresponding change. This information is permanently stored on the paper graph. Paper graphs are often provided in roll form. This allows for a continuous monitoring of the circuit value. A combination recorder and meter unit is illustrated in Figure 15-20.

Figure 15-20. A combination clamp-on meter and chart recorder provides both an accurate measurement and a permanent record of the values. (Courtesy Amprobe Instrument Co.)

Oscilloscope

Another visual readout device that has become commonplace in the electrical and electronics field is the oscilloscope. This device provides an electronic display of electrical waves. The display is shown on a cathode-ray tube. This tube operates in the same manner as the picture tube of a television set. The oscilloscope has three basic sections inside its case. There are a cathode-ray tube, a time-base generator, and a vertical amplifier. A coating on the inside of the cathode-ray tube converts electrical energy into light energy. A stream of electrons is created in the tube. This stream is directed toward the face of the tube. The coating on the inside of the tube glows in one spot when hit by the stream of electrons.

If there were no other sections of the oscilloscope, the beam would produce a spot of light in the center of the screen. A unit called a time-base generator provides horizontal motion to the beam. It controls the speed at which the beam moves across the face of the tube. Speeds range from about 1 second per sweep to less than 0.1 microsecond per sweep. The time rate of this sweep can be varied in order to synchronize the time-base frequency with that of the observed waveform.

The other major section of the oscilloscope is called the vertical amplifier. This section provides vertical movement to the waveform being observed. The interrelation of the vertical information and the time-base generator allow the electron beam to move to any place on the face of the cathode-ray tube. The beam normally moves at a very high rate of speed and appears as a solid, connected line rather than a single dot. The oscilloscope normally does not have the capability of storing or recording the visual image it creates. An oscilloscope similar to those used in industry or the laboratory is shown in Figure 15-21.

Figure 15-21.   The oscilloscope displays a visual image of the electrical wave with respect to time. (Courtesy Simpson Electric Co.)

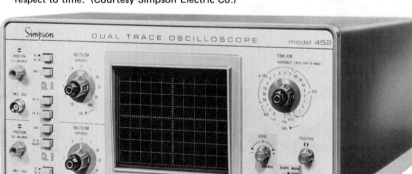

SUMMARY

Knowledge of test and measurement equipment is essential for persons working in electrical occupations. These persons need to know how the equipment functions and how to use it if they are to be successful in their work.

Both analog readout and digital readout test equipment are in use today. Analog equipment uses the principles of electromagnetic induction to convert a current flow into a rotary motion. The rotary motion of the meter movement provides a reading on the calibrated scale of the meter. The operation of the meter movement is similar to that of an electric motor. Both operate on the principles of repulsion and attraction of magnetic fields.

Limitations for analog meters include the amount of current in the armature, the direction of current flow, and voltage developed across the meter movement. The basic meter movement circuit is modified in order to measure other quantities. These would include resistance and temperature measurements.

Meters that measure voltage, current, and resistance are called multimeters. These units have the capability of extending the range of measurement from a low value to a much higher value. This is accomplished by means of a selector switch and some additional components. The function of the meter can also be selected by the use of a second switching circuit in the meter. One limitation of the multimeter is the loading effect. The meter's resistance when it is equal to, or lower than, the resistance in the circuit it is measuring will often give a false reading. This is due to the additional resistance of the meter circuitry being included in the circuit as the measurement is being made. This is only true for voltage measurements.

One method of overcoming this limitation is to use a meter whose resistance to the measured circuit is always a high value. This is done with an electronic voltmeter. The electronic voltmeter has an input resistance of at least 1 megohm on each of its ranges. This value is high enough to not load the circuit.

Electronic voltmeters use either an analog meter movement or a digital display. These are used to provide a readout of the values being measured. Both of these types of measuring devices require a power source to operate their internal circuitry.

Meters are connected into the circuit in different ways. These depend upon what is to be measured. A voltmeter's leads are connected between two points in a circuit. Voltage is a difference in potential between two points in a circuit. The leads from the voltmeter are connected to these points when a measurement is made.

Current is measured by wiring the meter into the circuit. This requires a physical break in the circuit. Another way of measuring current is to use a transformer type of meter. These are called clamp-on ammeters. They only work when current is changing in the circuit being tested. The theories of transformer action apply for use of this type of meter. A third method of measuring circuit current is to measure the voltage drop across a specific value of resistance in the circuit. The voltage drop and the resistance value are used with Ohm's law to calculate the current flow.

Resistance measurements require the removal of the component from the circuit. If this is not done, a false reading may be obtained. This is particularly true if other components are connected across the specific component whose resistance is being measured.

When making connections to the circuit for a meter reading, it is a wise idea to turn off circuit power. This is true even for voltage readings.

Two visual readout devices in common use are the graph recorder and the oscilloscope. Both give an indication of circuit values as they relate to time. No other measuring devices have the capability of showing a time relationship. The graph recorder uses a motor to move the graph paper or the marking pen. The quantities of circuit values are indicated by a vertical movement on the graph. Time is shown by horizontal movement.

The oscilloscope is an electronic device that displays the electrical waves on the face of a cathode-ray tube. A time-base generator in the oscilloscope provides horizontal motion for a beam of electrons. A vertical circuit provides vertical motion to the same beam. These two forces give the electron beam movement to any point on the face of the tube. The rate of movement is so rapid that the human eye observes this information as a solid line. The frequencies of the time-base generator and the observed electrical wave are synchronized to display one or two waveforms on the oscilloscope tube face. The oscilloscope has the advantage over the chart recorder of operating at much higher frequencies.

QUESTIONS

15-1. Devices used to measure circuit values are called

    a. meters

    b. quantum units

    c. measurers

    d. amplitude monitors

15-2. Measuring devices are limited by

    a. range

    b. polarity

    c. types of current

    d. all of the above

15-3. Current-measuring devices use ____ to extend their range.

    a. multipliers

    b. range extenders

    c. shunts

    d. bypass units

15-4. Voltage-measuring devices use ____ to extend their range.

    a. multipliers

    b. range extenders

    c. shunts

    d. bypass units

15-5. Electronic measuring units are called

    a. vacuum-tube voltmeters

    b. electronic multimeters

    c. transistorized multimeters

    d. all of the above

15-6. The voltage-measuring device is connected in ____ to the circuit to be measured.

    a. series

    b. parallel

    c. any way possible

    d. all of the above

15-7. The current-measuring device is connected in ____ to the circuit to be measured.

    a. series

    b. parallel

    c. any way possible

    d. all of the above

15-8. The resistance-measuring device is connected in ____ to the component or circuit to be measured.

    a. series

    b. parallel

    c. any way possible

    d. all of the above

15-9. Devices that measure values with respect to time are the

    a. time plotter

    b. graphic recorder

    c. oscilloscope

    d. answers b and c above

15-10. Resistance measurements are done

    a. with circuit power on

    b. with circuit power off

    c. with no relation to circuit power

    d. none of the above

16

Solid State

The evolution of solid-state devices has great significance in the electrical field. Solid-state devices such as diodes and transistors have provided the necessary components for miniaturization of control panels. Applications of these have resulted in more efficient devices and less of a power demand for control circuitry. The use of these devices permits faster and more accurate equipment operation. In addition, solid-state devices are used in place of relays in controllers. Mechanical failures are minimized because of the lack of moving parts.

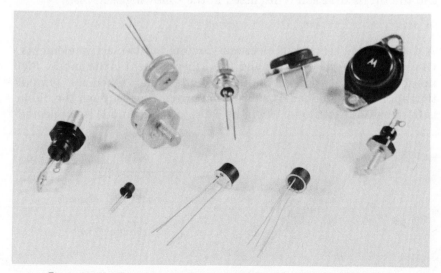

Figure 16–A. Semiconductor devices used in electrical and electronic units.

Persons employed in electrical occupations are being exposed to a variety of solid-state devices. The theories behind these devices and their applications in functional units are essential knowledge. This chapter is devoted to the understanding of the solid-state devices used today.

SEMICONDUCTORS

Material presented to this point in the book is related to the ease in which electrons flow in a circuit. The text material points out that resistance in a circuit is often an undesirable characteristic. This is not true in all situations. There is a family of devices that is designed to operate in a resistive condition. They are not true conductors nor are they true insulators. This group is categorized as semiconductors. Most semiconductor material is manufactured by adding a controlled amount of impurities to an insulator. The most common material used for this is silicon.

Semiconductor devices used in electrical applications have one feature that makes them greatly different from typical resistive devices. This is the semiconductor device's capability of having its internal resistance varied. The amount of resistance is controlled by an EMF applied to the device. Each general type of semiconductor has some very definite characteristics. These relate to how they operate as well as operational limitations. The most often used semiconductor devices are the diode, the transistor, the silicon-controlled rectifier, and the thyristor. Each is discussed in the following sections.

Diodes

A diode device as used in electrical circuits has two active elements. One of these is the cathode and the other element is the anode. Figure 16-1 shows two semiconductor diodes and a schematic symbol that applies to all diodes. The electron current flow through the diode is from cathode to anode. When the cathode of the diode is connected

Figure 16-1.   The semiconductor diode and its schematic symbol.

to the negative part of a circuit and the anode is connected to the positive part, the diode has a very low internal resistance. This is due to the voltage applied to the diode.

When the polarity of the applied voltage is reversed, the internal resistance of the diode becomes very high. These two conditions are shown in Figure 16-2.

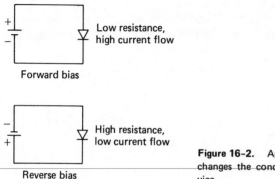

Low resistance, high current flow

Forward bias

High resistance, low current flow

Reverse bias

Figure 16-2. Applied voltage changes the condition of the device.

The circuit shown is not very practical. When the diode is forward biased, or in the low internal resistance state, its low internal resistance would allow so much current flow that the diode would probably burn up. This circuit is shown only as a means of illustrating the two states of the diode. A more practical circuit will include either a load or a current limiting resistor.

In many ways the diode action is similar to that of a check valve. The direction of current flow is limited to only one direction. The check valve does this same thing with fluids. This is illustrated in Figure 16-3. The diode uses electrical principles to shut off current flow when the current direction is opposite its normal flow path. The diode is in reality a voltage-controlled resistance in the circuit. The voltage applied across the elements of the diode determines the resistance state of the diode. Forward bias sets the conditions for "on" and a very low internal resistance. Reverse bias creates a high internal resistance and attempts to stop all current flow.

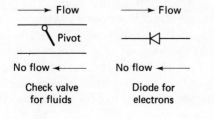

Flow   Flow

Pivot

No flow   No flow

Check valve   Diode for
for fluids   electrons

Figure 16-3. A diode acts as a check valve, allowing current flow in only one direction.

There are two general classifications for diodes. These are directly related to the type of work the diode does in an electric circuit. One group of diodes is called *signal* or *switching* diodes. These are relatively low power devices. They are used in logic and similar electric switching circuits. The second group is called *power* diodes. These are used as active devices in power circuits. A common name for this type of diode is a *rectifier* diode.

Switching diodes are found in the logic circuits that control other circuits. An example of a simple logic circuit is shown in Figure 16-4.

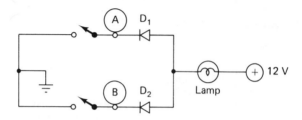

Figure 16-4.   Switching occurs by applying the correct bias to a diode.

This circuit shows two diodes, D_1 and D_2, whose anodes are connected to a lamp. The lamp is wired to the positive terminal of a power source. The cathode of each diode is connected to a switch, and the switch lead is wired to circuit common. If either of the switches are in the "on" position, the lamp will light. Current flow is from circuit common, through the "on" switch, the diode, the lamp, and to the positive terminal of the power source. Assume that the output terminal of an automatic machine controller is connected to point A in this circuit. When the controller is ready for operation, its output voltage is on the order of 1 volt. This is much lower than the +12 volts on the other side of the lamp. There is a major difference in voltage between point A and the +12 connection. A current will flow through diode D_1 and the lamp will light. If a second controller is connected to point B, this same set of conditions will still turn on the light. This type of circuit is called a *gate*. It controls another circuit. Specifically, this gate is called an OR gate. Either one OR the other input will turn on the output lamp. There are other types of logic gates, but their theory is not a part of this book. The purpose here is to show how a diode is used in a low-power switching circuit.

Power diodes may also be used for switching. The difference between the two types relates to the amount of power each type can safely handle. Power is the product of voltage and current in electrical circuits. Diodes are classified as either low power or high power. Low-power diodes used as switches are rated up to 500 milliwatts. This is a maximum voltage-current rating. High-power diodes are rated from 1 watt to over 120 kilowatts. The higher-powered diodes are rated in both operating voltage and current capabilities rather than a wattage rating. This provides a very concise set of circuit values for the diode.

Diode ratings for a specific device are obtained from the manufacturer's specifications. Each diode is assigned an industry type number. Diodes classified in this manner start with a 1N as a part of their number. The 1N is used as an identifier for diodes. There will be from two to four numbers after the letter N. For example, a diode using this classification system might be assigned the number 1N1486. This number is found in a master classification book. The book also provides operational information for the diode.

The information most often used with any diode relates to its operating voltage and current capabilities. The operational curve for a 1N4148 switching diode is shown in Figure 16-5. This type of curve is fairly accurate for diodes in general. The difference, of course, is in the specific voltage and current ratings for this diode. The graph shows both forward and reverse characteristics for the diode. The terms I_F and V_F refer to conditions when the diode is forward biased. Note the small amount of voltage (1 volt) forces a current flow of 10 milliamperes. In the reverse bias circuit a 75-volt pressure only forces a current flow of 5 microamperes. An increase of voltage will force a rapid increase of this current. The diode will go into a breakdown condition as the voltage reaches a 100-volt level. This is the point of destruction for this diode. This diode will not work safely at this

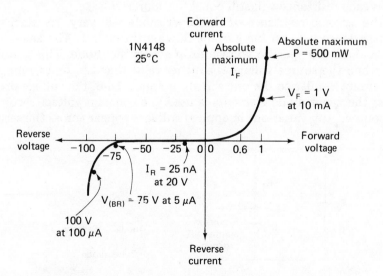

Figure 16-5. Operating characteristics for a switching diode.

high a reverse voltage. The two most often used specifications for a power diode are the forward current and the reverse voltage rating. Forward current refers to the maximum amount of safe current flow through the diode when it is forward biased. The reverse voltage rating is the maximum value of voltage that can be applied to the diode elements when the diode is reverse biased.

A third use for the diode is as a voltage regulator. The curve shown in Figure 16-5 shows how the current flow holds at a constant value and then drops very rapidly as the voltage changes from 75 to 100 volts. If the diode is placed in a circuit, this can be used to advantage. Every diode has this type of reverse bias curve. The values of voltage and current will vary from one diode to another, but each will exhibit this breakdown curve. The area where this occurs is called the *zener region* of the diode. A diode may be wired into a circuit to take advantage of this action.

If the diode is operated in the zener region, it becomes a voltage-regulating device. This is illustrated in Figure 16-6. A dc voltage is applied across the resistor and the diode. Note the symbol for the zener diode. This diode is placed in the circuit so that it is in a reversed bias condition. The voltage in the circuit forces a current flow. A specific voltage drop appears across the zener diode. The remaining voltage develops across the resistor. This is an application of Kirchhoff's voltage law. The sum of the voltage drops across each component in a series circuit will equal the voltage supplied by the source. The values of each resistance in the circuit may change, but the sum of the drops across each resistance will still equal the source voltage.

The internal resistance of the zener diode will vary. Its exact resistance depends upon the current flowing through it. The zener action maintains a specific voltage drop across the diode. This is only true when the source voltage is a higher value than the zener voltage. In a circuit similar to the one shown in Figure 16-6, the voltage drop across the zener diode elements is used as a constant voltage secondary source. Any variations in applied voltages appear across the series

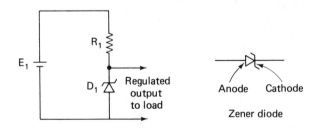

Figure 16–6.   A zener diode acts as a voltage regulator in the circuit.

resistor. Variations may occur owing to a change in the amplitude of the source voltage, or they may be due to changes in the value of the secondary load resistance as the device operates. The zener voltage level is maintained as long as the circuit current does not exceed the limitations of the zener diode.

One other group of diode devices that has recently been utilized to a great extent is the light-emitting diode (LED) family. This diode is made from gallium and phosphide rather than silicon. This device will produce a visible light as a current flows through it. The colors of the light are red, orange, green, and blue, depending on the exact chemical mixture used in the manufacture of the LED. In many cases a lens with a magnifying device is included as a part of the LED housing. Figure 16-7 shows the LED symbol and common commercial LED packages. These devices are used instead of filament bulbs as indicators on control panels.

Figure 16-7. The light-emitting diode (LED) and its schematic symbol.

Another application of the LED is as a numerical readout device. Typical LED readouts are shown in Figure 16-8. The readout is made up of a group of seven segments. Each segment has an individual set of LED's. Current from a source will light an appropriate segment. Any combination of segments may be lighted at one time. This causes any number from 0 through 9 to be displayed. LED readouts are used in computers, calculators, and machine control systems.

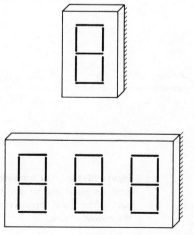

Figure 16-8. LEDs arranged as a seven-segment readout display show numeric values.

Transistor

Internal resistance and the resulting current flow in a diode are controlled by the amount of applied source voltage to the diode. This has some serious limitations in circuit work. As a result of research at Bell Telephone Laboratories, the transistor was developed. The transistor has three elements. These are the base, emitter, and collector. The transistor is produced with one of two polarities. One is when the collector element is positive, and the other is when the collector element is negative. The symbols for both types is shown in Figure 16-9. One of these is called an *NPN* transistor and the other is called a *PNP* transistor. The symbol for both of these is quite similar. The only difference is the direction of the arrowhead on the emitter element and the direction of current flow through the transistor.

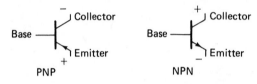

Figure 16-9. Bipolar transistors are constructed as either *NPN* or *PNP* devices.

In most transistor applications the transistor is said to have two circuits connected to its three elements. One circuit is the input, or control, circuit. It is usually connected to the emitter and base elements. Less than 5% of the current used by the transistor is used in the control circuit. The second circuit is the emitter and collector circuit. The load to be controlled is connected in this circuit between the transistor and the power source. The circuit connections are shown in Figure 16-10. The emitter and collector circuit has about 95% of the circuit current flowing in it during operation.

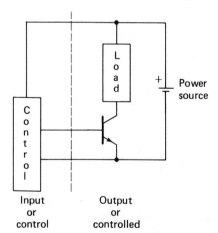

Figure 16-10. The transistor has both an input and an output circuit. The input circuit controls the output circuit.

In a practical transistor circuit a small amount of power operates the control circuit. This small amount of power is utilized to control much larger amounts of power in the output circuit. The transistor, therefore, is a device that operates like a controlled variable resistance in the circuit. A schematic diagram of this application is illustrated in Figure 16-11. The control portion of the transistor is often the base and emitter elements. The load, or controlled, part of the circuit is connected to the emitter and the collector elements of the transistor. This circuit configuration is called a common emitter circuit because the emitter element of the transistor is used in both the input and the output circuitry.

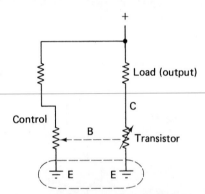

Figure 16-11. Application of the transistor as a control device.

The explanation of how the transistor functions is better understood when one realizes the derivation of the word "transistor." This name is composed from the phrase "transfer of resistance." Small variations of resistance in the input circuit control larger variations of resistance in the output, or load, circuit.

The transistor is able to act like a switch in a circuit. Each transistor has a set of operational characteristics. Some of these are directly related to the application of power to the input circuit. When power is applied to the input, or base-emitter, circuit the transistor's emitter-collector resistance is established. The application of the power requirements on the input circuit is referred to as settling the bias of this circuit. The term *bias* refers to a voltage or current applied to an element of a transistor or a vacuum tube that helps to establish operating conditions for the device.

The operational point of the transistor may be established so that the internal resistance of the transistor is so high that no current will flow between emitter and collector. This condition is called *cutoff*. Another operational point may reduce internal resistance in the transistor to such a low value that enormous quantities of current will

flow. This condition is called *saturation*. When the transistor is operated at these extremes, it will act as a switch in the circuit. This is
shown in Figure 16-12. The load and the emitter-collector elements
of the transistor are wired in series. Kirchhoff's law for a closed loop
is applied in this circuit. The specific application here includes the
load, the transistor elements, and the power source. The sum of the
voltage drops developed by a current flow through the load and the
transistor will equal the applied source voltage. The voltage drops in
a series circuit are directly proportional to the size of each resistance.
Based on these principles, the transistor's voltage drop is very low
when it is in a saturated state. If this is the situation, then the voltage
drop across the load is high. Power is then applied to the load for
operational purposes.

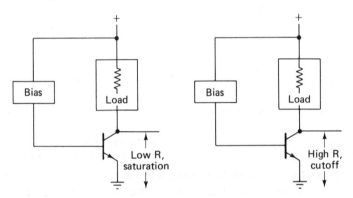

Figure 16-12. A transistor may act as a switch when it is connected to be either
fully on or fully off.

When the transistor is in a state of saturation, its internal resistance is very high. Normal relationships between resistance of the load
and transistor will possibly allow a ratio of at least 10:1 between saturation and load resistances. In this condition the greatest voltage
drop in the circuit is across the transistor elements. The power available to operate the load is very low under these conditions. The load,
therefore, is in an "off" condition. All this is accomplished when the
bias on the control element of the transistor is changed. The bias is
often a dc voltage. The source of this dc voltage will often be a con-
trol-type circuit. It may be the output of a microprocessor device or
computer. In any case, it is a very low power value that sets the operational conditions of the transistor. When the transistor is operated
and changes rapidly from saturation to cutoff, it acts as a switch in
the circuit.

Another manner in which the transistor operates is when it acts as an amplifying device. The definition of an amplifier in electrical terms is a device that uses a small amount of power to control a larger amount of power. The base-emitter and emitter-collector circuits of the transistor illustrate this very well. The schematic diagram of Figure 16-13 will be used as a basis for showing this. The circuit has a 100-kilohm resistance, which is used to establish the operational point of the circuit. A 5-kilohm load is series wired to the transistor emitter-collector circuit. The circle with the letters "in" represents a voltage that is added to the circuit after the operating bias is established. This additional voltage will change the operating point of the circuit. These changes are reflected in the amount of power consumed by the input circuit.

Under normal amplifier conditions, the operating point is midway between saturation and cutoff points. This point is at the 1 volt E-B level for this circuit. Ten microwatts of input power is used to establish this condition. The balance of the information on the same line in the chart relates to the output circuit. The resistance of the emitter-collector is equal to that of the load at this point. The output circuit current is the same through both the load and the transistor since they are series connected. The output power is 22 milliwatts under these conditions. The power used by the load is equal to the power used by the transistor at this point. The gain, or amount of amplification, is determined by this formula:

$$\text{gain} = \frac{\text{output}}{\text{input}}$$

In this circuit the power gain at this mid-point is 2,200. This means that the input power action sets the conditions to control 2,200 times its own power-handling capability. This figure will vary from one transistor type to another. It also may change when using different circuit values.

An additional set of data shows what happens as the input power changes. When the input power increases to 16.9 microwatts, the power in the transistor drops to 13.2 milliwatts. At the same time the power used by the load increases to 52.8 milliwatts. This small change of 6.9 microwatts in the power used by the input circuit has controlled a change of 30.8 milliwatts in the output circuit.

The changes illustrated use a small power device as an example of how amplification occurs in the transistor. Other values of circuit components function in the same manner to control larger amounts

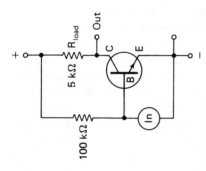

	Input Conditions			Output Conditions				
E·B Volts	E·B Current	E·B Power	E·C Volts	Load Volts	E·C Current	E·C Power	Load Power
1.5	15 μA	22.5 μW	0	20	4.0 μA	0.0 mW	80.0 mW
1.4	14 μA	19.6 μW	2	18	3.6 μA	7.2 mW	64.9 mW
1.3	13 μA	16.9 μW	4	16	3.3 μA	13.2 mW	52.8 mW
1.2	12 μA	14.4 μW	6	14	2.8 μA	16.8 mW	39.2 mW
1.1	11 μA	12.1 μW	8	12	2.4 μA	19.2 mW	28.8 mW
1.0	10 μA	10.0 μW	10	10	2.2 μA	22.0 mW	22.0 mW
0.9	9 μA	8.1 μW	12	8	1.8 μA	21.6 mW	14.4 mW
0.8	8 μA	6.4 μW	14	6	1.4 μA	19.6 mW	8.4 mW
0.7	7 μA	4.9 μW	16	4	1.0 μA	16.0 mW	4.0 mW
0.6	6 μA	3.6 μW	18	2	0.5 μA	9.0 mW	1.0 mW
0.5	5 μA	2.5 μW	20	0	0.0 μA	0.0 mW	0.0 mW

Figure 16-13. Various values of the input bias control the output values. These are illustrated in this chart.

of power. Circuits are in use that may control over 500 watts of power. Actual values for control and controlled circuits vary. These are selected when the circuit is designed.

Transistors are often used with a varying voltage rather than a dc voltage at the input to the control elements. With a varying voltage the transistor will amplify the shape of the electrical input wave. Devices used in this manner are called *signal amplifiers*. A schematic diagram for a signal amplifier is shown in Figure 16-14. A small ac signal is connected between base and emitter of the transistor. This is considered to be the input signal. The bias on these elements is set so that the transistor is at its mid-point of operation. The signal voltage changes the current flow in the control circuit. This, in turn, changes resistance of the emitter-collector circuit. The output connections of this circuit are connected between emitter and collector. The input signal voltage varies the resistance of the output circuit in this manner. Information given in Figure 16-13 will verify this action.

A decrease in resistance between the emitter-collector element will drop the voltage across these elements. The transistor is in series with a resistor. The voltage increases across the series resistor as the voltage drops across the transistor. The opposite is true as the resistance increases across the transistor output connections. The result is a larger inverted wave at the output. The output wave has the same basic form as the input wave. This format is valid for many amplifier circuits.

Transistors are identified by a set of standard numbers. These numbers usually start with the prefix 2N and follow with from two to four numbers. A typical transistor number using this system is 2N3066. Do not be too surprised when you find semiconductors whose identifying numbers differ from the industry standard numbering system. Many semiconductor manufacturers have a set of "house numbers." These are used to identify either special design devices or

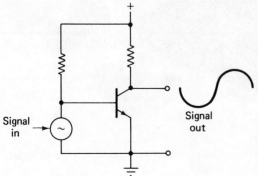

Figure 16-14. A signal amplifier uses a small amount of power to control a larger amount of power. Signal inversion occurs in this type of circuit.

those which have not been assigned a number from the standard series. In some instances the equipment manufacturer will purchase
semiconductors that have had the equipment manufacturers' part
number printed on the case rather than a standard number. Propri-
etary numbers such as these may often be cross-referenced to industry
standard numbers when a replacement is needed.

There are several transistor specifications listed in data sheets for
each number semiconductor. Certain of these are more important
than others. These include current gain, power dissipation, and breakdown voltage. Current gain is often called *beta* or h_{fe} of the transistor.
This figure is a ratio of output current to input current. It shows the
amount of current amplification in a specific transistor design. Maximum power dissipation relates to the amount of power the transistor
wastes by generating heat as it conducts. If the transistor uses power
in addition to the rated value, it will burn up. This figure is of great
importance in design work. The third item of concern is the breakdown voltage rating. This refers to the maximum voltage that may be
applied between emitter and collector without having the transistor
fail. If a higher-than-rated voltage is applied to these elements, an excess of current will flow. This leads to a condition that will generate
additional heat in the semiconductor. The additional heat leads to
more current flow. This continues to build up until the semiconductor is destroyed.

One way of keeping a semiconductor device cool is to mount it
on a heat-radiating device. These devices, pictured in Figure 16-15,
are called *heat sinks*. They are produced in a variety of sizes and

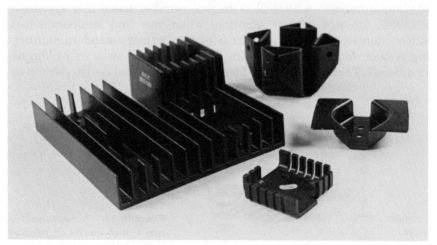

Figure 16-15. Heat-radiating devices, called heat sinks, are used to dissipate excess
heat and keep the device cool.

shapes. Each heat sink is designed with a specific area of radiation. The additional radiation area allows the heat buildup in the semiconductor to dissipate into the air. The result is a cooler operational level for the semiconductor. In extreme cases a fan is used in addition to the heat sink. The fan keeps the air moving past the heat sink. It permits higher semiconductor temperatures than would be present if there were no heat-dissipating devices.

Thyristors

Another family of semiconductor devices is the thyristor family. These include silicon-controlled rectifiers and triacs. These devices operate in a manner similar to the diode. They have special features that make them unique and readily adaptable to power control circuitry. Thyristors are three-terminal devices. This makes them similar to transistors. Thyristors have both control elements and controlled elements. The major difference between a thyristor and a transistor is that the thyristor does not require any control action once it is turned on. The result is a simple control circuit. The control circuit does not require power once it has completed its function.

Silicon-controlled rectifier (SCR). One member of the thyristor family is the SCR. This device functions in a manner that is similar to a rectifier diode. The rectifier diode permits current flow in one direction in the circuit. This is illustrated in Figure 16-2. An ac voltage is applied to the circuit. The diode action switches current on in the load during one portion of the input wave cycle. This portion is when the applied voltage to the diode forward biases it. When this voltage reverse biases the diode, it does not permit current flow in the circuit. The result of forward biasing the diode allows a current flow through the load. A voltage is developed across the load. Power is consumed in the load. Work is done.

The amplitude of the voltage developed across the load is directly proportional to the amplitude of the ac wave of the source. The only control of current flow is the amount of voltage at the source and the value of the load resistance. The amount of power used by the load cannot be conveniently controlled until the voltage or current is changed. A reduced amount of either of these values results in less power at the load. The ability to perform work is also reduced under these conditions.

These limitations are overcome with the SCR. A picture of SCR devices and their respective schematic symbols are shown in Figure 16-16. Current flow in the SCR does not occur until the device is turned on. When a circuit that has a SCR is first turned on, the de-

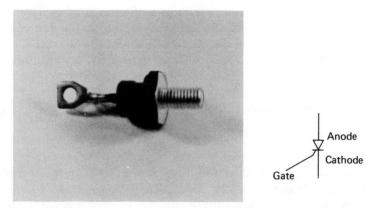

Figure 16-16. Silicon-controlled rectificer (SCR) and its schematic symbol.

vice will not allow a current flow in either direction. This condition remains until a signal or triggering current is applied to the gate control element. When the gate is triggered, the SCR's anode-cathode resistance is reduced to a very low value. Conduction from cathode to anode occurs. This condition continues even though the gate current is turned off. It will continue until the cathode-anode voltage is removed or the polarity of this voltage is reversed.

An application of the operation of the SCR in a circuit is illustrated in Figure 16-17. A SCR is series connected to a load. The circuit is completed by connecting an ac source across the SCR and load. The trigger circuit is also connected between the SCR gate and the ac source. In this example the trigger circuit is set at its mid-point. When the current in the trigger circuit reaches a predetermined value, it will turn on the SCR. In this example the trigger wave has to reach its maximum value before the SCR will conduct.

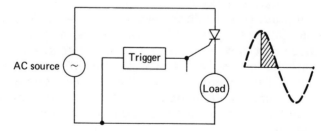

Figure 16-17. Application of the SCR in a circuit. This device only works with DC-controlled loads.

The point of conduction may be varied to almost any point on the first quarter of the wave's cycle. This is illustrated in Figure 16-18. This represents input and output waves in the circuit. The letters *A*, *B*, and *C* represent settings of the trigger circuit. Point *A* on the input wave drawing relates to the waveform shown on the output *A* wave. The same is true for points *B* and *C*. The timing of the input wave selects the timing of the output wave.

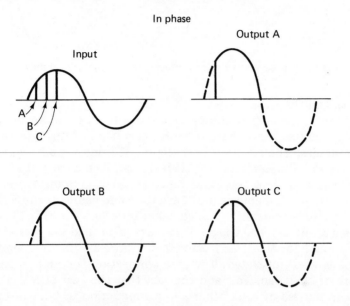

Figure 16-18. Operation of the SCR when gate and anode voltages are in phase is limited to the first quarter of the cycle.

This illustration is true when both the input and the output waves are in phase with each other. Maximum power in the output circuit is at the mid-point of the wave. This type of circuit maintains full power to the load during that portion of the time period when the SCR is conducting. If the input wave is out of phase from that of the output wave, the point of control may be shifted to some other portion of the "on" cycle. This is illustrated in Figure 16-19. A phase shift allows any portion of the first half-cycle of the wave to be rectified. When this phase shifting circuit is a part of the trigger circuit, there is an additional area of control time. The concept of full power for the "on" time is still a valid consideration in this type of circuit.

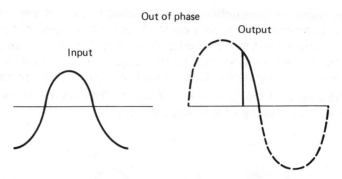

Figure 16-19. Out-of-phase operation permits use of the SCR over all the first half-cycle.

Triac. One limitation of the SCR is that it only operates during half of the ac cycle. This limits the control function to loads that require dc sources. Many devices requiring control of this type operate on ac voltages. The addition of a second SCR in the control package will overcome this limitation. The device is called a triac. It has three elements just as does the SCR. Since this device operates in ac circuits, the power elements cannot have the same names as the SCR elements. These elements are called *main terminals*, or MT. There are two main terminals. The trigger element is still called the gate. A schematic symbol for the triac is shown in Figure 16-20. An operational concept is also illustrated. The gate connection is common to both sections of the triac. This and the fact that the two SCR's are connected in reverse of each other allow conduction in both directions in the circuit. The critical thing here, also, relates to the "on" time of the device. The output or controlled wave occurs on both the positive and the negative segments of the ac wave. This permits operation from an ac source and an ac-operated load. It also doubles the effi-

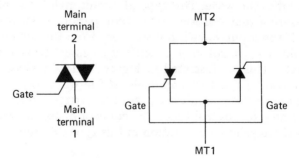

Figure 16-20. A triac operates on AC circuits. It is basically two SCRs connected back to back in one package.

ciency of the triac compared to the SCR. The concept is illustrated in Figure 16-21. The output of the circuit is still an ac wave under these conditions. The difference lies in the amount of time the circuit is on.

Thyristor devices have two critical ratings. One has to do with the current-carrying ability of the device. The second refers to the level of voltage that is applied to the thyristor. Both of these ratings, if exceeded, will cause destruction of the triac.

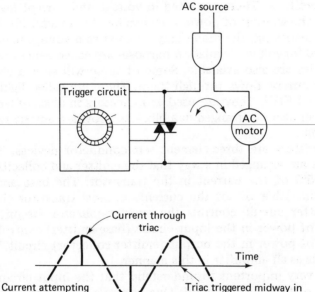

Figure 16-21. The triac as used to control an AC circuit load.

SUMMARY

Solid-state devices are an essential part of the electrical field. These devices include the diode, transistor, silicon-controlled rectifiers, and thyristors. Each of these is produced from a semiconductor material. The most used semiconductor material has a silicon base.

A diode is a two-element semiconductor. It acts like a two-state resistance when connected in a circuit. The state of the diode is dependent upon the polarity of the voltage applied to its elements. In one state it is considered to be forward biased. This is its "on" or low-resistance condition. When the applied voltage is reversed the diode is reverse biased or "off." This is its high-resistance condition. The diode, therefore, acts as a voltage polarity controlled switch in a circuit. Two essential characteristics of all diodes relate to the amount of current the diode will safely handle when it is forward biased and the level of voltage that can be applied to the diode when it is reverse biased.

Diodes are classified as either signal or power diodes. Low-power diodes are considered signal or switching diodes. These usually can handle up to 500 milliwatts of power. The power diodes are also called rectifiers. These are rated in voltage and current levels rather than by the amount of power they can handle. In addition to switching and rectifying, the diode may be used as a voltage regulator. Diodes used for voltage-regulation purposes are called zener diodes. Special diodes are also avialable. Some of these will emit a visible light while a current flows through them. These are called light-emitting diodes or LED's. They are used as indicators in place of lamp bulbs. LED's can also be arranged in a block to provide numeric readout information.

Transistors are three-element semiconductor devices. The three elements are arranged in a way that the emitter and collector handles about 95% of the current in the transistor. The base and emitter handle the other 5% of the current. In most transistor circuits the base-emitter circuit controls the emitter-collector circuit. A small amount of power in the input circuit (base-emitter) controls a larger amount of power in the output (emitter-collector) circuit. The transistor acts as an amplifier in this manner.

It is very important to understand that the input circuit is independent of the output circuit. One *controls* the action of the other. The input circuit power is not transferred through the transistor into its output circuit. Again, one acts as a control for the other. This concept is basic to understanding how an electronic amplifier functions.

Transistors may be operated in either of two extreme conditions. One of these is called saturation. This condition is when the internal resistance is very low. Large amounts of current flow through the transistor at this time. The other condition is called cutoff. The internal resistance of the transistor is very high in this condition and little or no current flows. The transistor may also be operated somewhere between these extreme conditions.

All conductive devices may be assigned an industry number. Diodes registered in this manner use a 1N . . . number sequence. Transistors use a 2N . . . sequence for their numbers.

Specifications for transistors include current gain, power dissipation, and breakdown voltage. Current gain, or beta, is the ratio of output current to input current in an operational circuit. Power dissipation refers to the amount of heat the transistor will withstand while it is conducting. The third rating refers to the level of applied voltage that can safely be applied to the transistor. Semiconductors are often mounted on a heat-dissipating device to help keep them cool. This device is called a heat sink. It radiates heat into the air.

Thyristors are another group of three-element semiconductors. Thyristors do not amplify. They are used to control a portion of an electrical wave. The silicon-controlled rectifier (SCR) is one type of thyristor. It acts in a manner similar to a diode except that it has one extra element. This element is the gate. A small current on the gate turns on the SCR. Once on, it stays on even though the gate current is removed. Turn-off is accomplished either by removing the applied voltage or reversing its polarity. SCR's work to provide dc operating voltages from an ac source.

Triacs are similar to SCR's except they can work in an ac circuit. The triac is made from two SCR's wired in a back-to-back connection. The triac can control an ac circuit. It works in a manner similar to that of the SCR. Thyristors have a maximum current rating and a maximum voltage rating.

All these devices are used to control power in electrical circuits. When the power is applied, work is accomplished. Control of the amount of power available to the load controls the level of work being accomplished.

QUESTIONS

16-1. A semiconductor has a low resistance when it is

    a. reverse biased

    b. forward biased

    c. out of the circuit

    d. independent of supply voltages

16-2. A forward biased semiconductor used as a switch will

    a. permit large amounts of current flow

    b. stop all current flow

    c. permit small amounts of current flow

    d. have no relation to current flow

16-3. Diodes are rated

a. as either signal or power diodes

b. with forward current vlaues

c. with reverse voltage values

d. all of the above

16-4. Diodes used as voltage regulators are called

a. zener diodes

b. regulator diodes

c. voltage-holding diodes

d. all of the above

16-5. The transistor is

a. able to act as a switch

b. three element device

c. able to use one pair of elements to control a second pair of elements

d. all of the above

16-6. Saturation is a ____ condition

a. full off operational

b. full on operational

c. mid-point operational

d. variable point operational

16-7. A device used to keep semiconductors cool is the

a. semiconductor cooler

b. radiator

c. heat sink

d. cooling fin

16-8. A diode that may be used as a triggered dc switch is the

a. triac

b. silicon transistor

c. silicon-controlled rectifier

d. switching diode

16-9. A device that may be used as a triggered ac switch is the

a. triac

b. silicon transistor

c. silicon-controlled rectifier

d. switching diode

16-10. Triggered diodes are used because they

    a. can control power

    b. can control current

    c. can control voltage

    d. have timing regulation

17

Production and Distribution of Electric Power

Electricity is produced in a variety of ways. Once produced, the electric power has to be transported from the generating plant to the user. The method of transportation and distribution is of importance to people working with electricity. The principles involved in production are basic electrical principles. Methods of transporting the electrical energy from the generating plant to the ultimate consumer follow basic electrical principles also. This chapter covers both the production and the distribution of electric energy as accomplished by electric power producers.

PRODUCTION OF ELECTRIC POWER

All commercial generation of electric power in the United States is alternating current. The frequency of this power is 60 hertz. Generating plants use ac generators called alternators. These alternators have a stationary armature and a rotating field winding. This is a reversal of the normal generator armature and field placement. The alternator has three sets of windings. It produces a three-phase alternating current. Each of the sets of coils in a three-phase alternator is spaced 120 degrees apart from the other coils. This produces three voltages that are related as illustrated in Figure 17-1.

Each set of coils in the alternator has two wires connected to it. Each pair of wires is connected to its load. The overlapping of the voltages produced in this three-phase system leads to some interest-

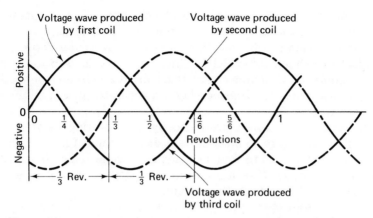

Figure 17-1. Three-phase AC as produced by the alternator of a power plant.

ing observations. For example, the sum of the currents at any instant of time in three of the six wires connected to the coils is zero. This is because some of the currents are moving away from the alternator into the load at the same instant as others are moving toward the alternator. If three of the six wires have a zero current flow, it is possible to eliminate them from the circuit and still have a functioning alternator. The result is a three-wire, three-phase system.

Two types of connections using three wires are illustrated in Figure 17-2. In actual practice all six wires from the alternator coils are used inside the machine. Three wires are all that are required to utilize this power. One arrangement for connecting the alternator coils is called a *wye* connection. The three coils are wired with one common

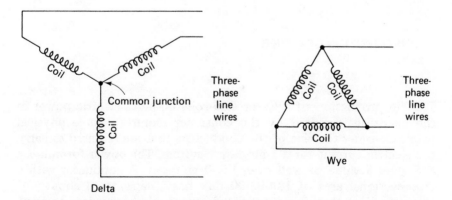

Figure 17-2. Delta and wye connections are used to connect alternator windings in a most efficient manner.

connection. The configuration looks like the letter Y, hence its name. The other connection is the *delta* connection. The three coils are connected in such a way that they appear in triangular form, which looks like the Greek letter delta. Thus this name is used to describe the triangular wiring connection. Each of these circuit connections requires three wires for transporting the generated power to the ultimate consumer.

The amount of power produced by a generating plant is quite high. This is necessary to meet present-day demands for electrical energy. It is not unusual to find electric power generating plants with a capacity of 1 megawatt or more. Actual capacity of the plant will depend upon the demands on the system.

There are some very important reasons to use both a three-phase alternator and the Y connection of the alternator coils. When a three-phase system is used instead of a single-phase system, the efficiency of the generating plant is much higher. In addition, the rotation of the moving coils is smoother owing to the three sets of coils connected to their respective loads. A third factor is that the wire required to carry three-phase power is about three quarters of the size required to carry a comparable single-phase power load. The reduction of wire size offers relatively smaller physical size of motors and transformers used in the three-phase system. These factors are important when one considers the many miles of wire required in both rotating machinery and conductors to transport the power from the generating plant to the consumer. Smaller wire sizes for comparable conductors mean that the support poles for the wires can also be of a lighter construction.

DISTRIBUTION OF POWER

The production of 120- or 240-volt, 60-hertz electric power in the quantities required by the consumer requires a large physical plant. Consider the size of the conductors that are required to carry the electric current for a 1-megawatt system. The power formula $I = P/E$ gives a value of well over 4,000 amperes. A conductor with a cross-sectional area of 1,000,000 mils has a capacity of about 770 amperes. More than five wires of this size would be required to carry this amount of current. The size of this wire is about 2-1/2 inches. This would require many closely spaced support poles to support the heavy wires that bring the current from the generating plant to the consumer. In addition, the large size of the conductors required in the alternator would make it a mechanical monster.

Another reason for transmission of high voltages at low current levels has to do with line losses. Electrical power is the product of voltage and current. The power formula is also written as $P = I^2R$. This is true when $I \times R$ is substituted for E in the original formula. When one considers this formula, it is reasonably easy to see that the power loss due to resistance in a line or wire is increased to a great extent as the current rises. The best way to reduce these losses is to operate the system with the lowest possible current value. This is accomplished with the high-voltage, low-current system in current use.

This system requires the use of a generator with an output voltage that is much higher than 240 volts. A common voltage for an alternator is 13,800 volts. Current demands for 1 megawatt of power with a voltage of this value are slightly more than 72 amperes. The size of wire required to carry this amount of current is slightly less than 17,000 circular mils, or $\frac{1}{8}$ inch. This higher voltage operation has reduced the cross-sectional area by more than 400 times. The power-handling capacity has not changed. The raise in operating voltage and corresponding drop in current have enabled the generating plant to be a more practical and less expensive size. The concept of the physical sizes of the turbines and alternators required to produce 240-volt power is almost beyond comprehension.

This variation of the voltage and current ratios is carried even further when the electric power is transported to the consumer. A typical system for production and transportation of electric power is shown in Figure 17-3. The 13,800 generated voltage is stepped up to a much higher value for transportation. This value is 345 kilovolts, for a step-up ratio of 25 to 1. High-tension lines are used to carry the power from the generating plant, across the country, to a consumer. If the consumer is a large industrial plant, the plant will often have its own voltage step-down system. This is called a *substation*.

Once the high voltage reaches its destination, it is reduced to an even lower voltage. The power being transmitted is essentially the same amount as that produced at the generating plant. Stepping the voltage down while maintaining the power level provides a higher level of current. Use of Watt's law will support this. Power is the product of voltage and current. If the voltage decreases, the current

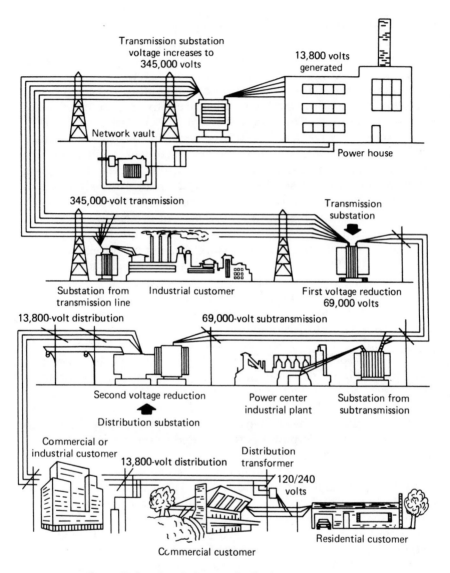

Figure 17-3.   A typical power distribution system map.

increases if the level of power is held constant. This is shown in Figure 17-4.

The 345-kilovolt level of voltage is reduced to 69 kilovolts at the substation. This second value of voltage is transported to a second substation. The second substation also steps the voltage down to a lower value. This new value is 13.8 kilovolts. Both the 69- and the 13.8-kilovolt voltages may be utilized by industry at that point. The

Voltage	Current	Power available
13.8 kV	72.5 A	≅ 1 MW
34.5 kV	2.8 A	1 MW
69 kV	14.5 A	1 MW
13.8 kV	72.5 A	1 MW
240 V	4167 A	1 MW

Figure 17-4. How voltage and current are changed for transportation purposes. This is accomplished without any major losses of power in the system.

industry would have to have its own substation system to provide more practical operating voltages in the plant.

The final reduction occurs in the community where the power is to be used. Additional step-down transformers reduce the 13.8 kilovolts to 240 volts. This voltage is distributed to homes and small commercial consumers.

The amount of power being transmitted remains at an almost constant value as these changes in voltage occur. If one can assume a close to 100% efficiency rate at each step of the distribution system, the resulting changes in current can be observed. The generating plant used as a model for this presentation produces 1 megawatt of power. Using 1 megawatt as a base for discussion, the current at the plant is just over 72 amperes. When the voltage is increased to 345 kilovolts, the current drops to less than 3 amperes. As the voltage continues to drop at each substation, the current increases. At the lowest voltage the current has risen to over 4,000 amperes. When this high current is used, it is divided among several consumers. The result is several small-sized feeder lines. Each line carries some of the current from the local distribution point to the consumer.

The system described is one form of distribution system. Its purpose is to carry large amounts of power from the generating plant to many consumers. It is very probable that a practical system will have more than one branch for power distributuion at each substation. There often are additional steps in voltage reduction between the 13.8-kilovolt and the 240-volt level. One system in common use uses several substations. Each substation will carry three-phase power at either 2,400, 4,160, or 13,800 volts to a distribution center. At the distribution center the power is connected onto the desired single-phase or three-phase main lines. The three-phase main lines connect the center to industrial or commercial uses. The single-phase lines connect to a step-down transformer that converts the voltage to a 240-volt level for use in the home or by small commercial and industrial customers. This last step reduces the single-phase voltage from 2,400 volts to 240 volts. A graphic representation of this part of the system is shown in Figure 17-5.

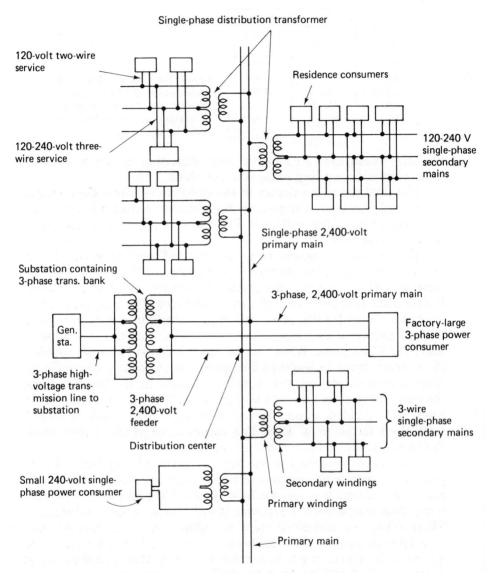

Figure 17-5.   A distribution schematic for electric power.

When the 240-volt service is connected to the home or small commercial area, it is delivered as a three-wire, single-phase voltage. The *service*, as it is called, is illustrated in Figure 17-6. Three wires carry the power from the pole to the building. Two of these wires are connected to the "hot" windings of a transformer. The third wire is called the *neutral* wire. It is connected to ground. Inside the building is another distribution system. This system differs from others in that it

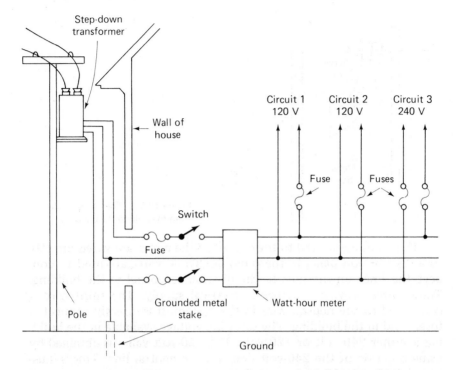

Figure 17-6. Distribution of electric power from local line to home or business.

does not use a transformer. This distribution system is often called a *service box*. Circuits are arranged in the box so that either 240 volts or 120 volts may be used. The 240 volts are obtained when connecting to each of the hot wires. A 120-volt source is obtained by connecting to one hot wire and the neutral wire. Circuit breakers or fuses are installed in these lines to protect the building from destruction in case of an overload condition.

SERVICE DISTRIBUTION

This section of the chapter deals with the distribution system found in the home or a small commercial establishment. It is that portion of the electrical distribution system most often seen by the consumer. The material in this section starts with the *service drop*. This term applies to the wires connecting the support pole and the building in which the power is to be used. A typical service drop is shown in Figure 17-7. Three wires connect the lines on the pole to the input terminals of the electric meter. In some installations these wires are run underground and are not seen until they connect to the input of the meter.

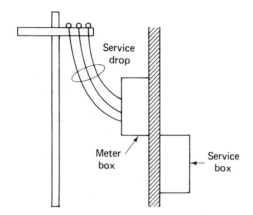

Figure 17-7. The service drop from the pole to the building.

The wires connected to the output of the meter are wired directly to a distribution panel in the building. Three wires are used to connect the lines on the pole to the distribution panel in the building. These wires provide a 240-volt electrical supply. The third wire is connected to the neutral wire in the system. It is directly connected to ground in the building. The resulting voltage available in the building is either 240 volt or 120 volt. The 120-volt value is obtained by using one side of the 240-volt line and the neutral line. This is illustrated in Figure 17-8. An installation found in the home or a small commercial user will have the option of using either of these voltages. In many cases both values are used. The 120-volt values are used for very light loads. These include radios, lamps, and small appliances rated at 20 amperes or less. Devices that consume larger amounts of power are operated from the 240-volt lines.

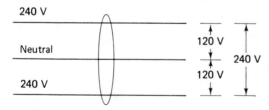

Figure 17-8. A three-wire 120-240 volt system uses this configuration.

The physical layout of the distribution panel is shown in Figure 17-9. The three wires from the meter are brought into the panel. Each of the two 240-volt wires are connected to a fuse or circuit breaker. These are the master units in this panel. Turning these off turns off the entire panel. The output side of each of these units is connected to a heavy piece of flat metal called a *bus bar*. The bus bar is a feeder for the individual circuits that fan out from the distribution panel to the loads.

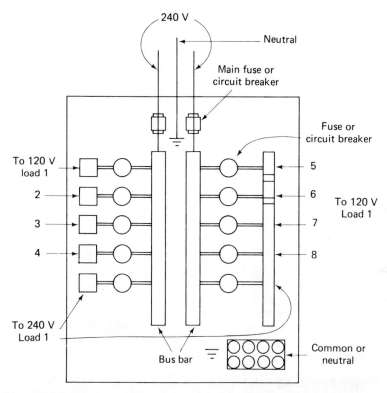

Figure 17-9. The distribution panel in the home or business uses fuses or circuit breakers as protective devices.

Several fuses or circuit breakers are connected to the bus bar. The output side of each of these units is wired to various lamps or wall outlets as required. The common, or neutral, line to each load is connected to the common, or neutral, connection in the panel. This is illustrated in Figure 17-10. When a 240-volt source is required for an air conditioner or electric stove, a different connection is required. These appliances require a direct wiring connection to each 240-volt source wire. A fuse or circuit breaker is connected to each of the two bus bars. One of the wires from the appliance is connected to each of these devices. This provides the required 240-volt source for the appliance.

Each circuit protection device is selected with the load in mind. Circuit protection depends on this. Typical ratings for these devices are either 15 or 20 amperes when the load is connected across a 120-volt source. Larger values of current as demanded by 240-volt loads will require units rated at 20 to 50 amperes. Information provided by the device manufacturer will identify the proper rating for circuit protection.

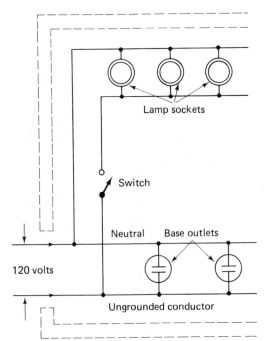

Figure 17-10. A single circuit wiring diagram in the home.

THREE PHASE ADVANTAGES

It is logical to ask about the reasons for using three-phase generating systems instead of single-phase power. One advantage is the comparable size of related components. Single-phase transformers and motors are larger than equal-power three-phase units. Another advantage is the ability to produce larger amounts of power in the three-phase system. This is because of the relationship of the three phases. When a load is connected to one phase of the three-phase system, its current capability is limited to that of the specific windings of the generator. Connections between windings of two phases produce a current that is 1.73 times larger than the single-phase connection. This is due to the interaction of the currents between each phase in the system.

It is possible to transmit three-phase power to a distribution station and then to connect a single-phase transformer to each phase winding on a three-phase transformer. The end result is the capability of having single-phase power to transmit to the consumer. A schematic for this type of system is shown in Figure 17-11. Each single-phase transformer is connected to one winding of the alternator There is one additional wire used in this system. This wire, which is connected to a center connection of the alternator, is called the *neu-*

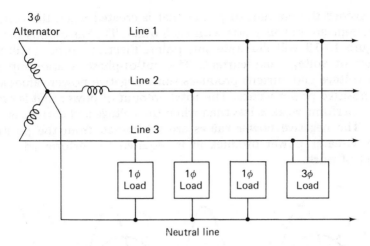

Figure 17-11. A three-phase wye connection distribution diagram.

tral wire. It is only used in a wye connection on the alternator circuit. This circuit has the advantage of being usable if one wire from the alternator should accidently be short circuited to ground. If this should occur, the balance of the circuit is still usable. In the delta circuit connection, a short circuit of this type would disable the entire system.

POWER FACTOR

The efficiency of any electrical source is related to its ability to transfer energy to the load. When an electric load is resistive in nature, maximum power transfer may occur. Both voltage and current are in phase in a resistive load. The formula for electric power is $P = E \times I$. When voltage and current are in phase, as shown in Figure 17-12, there is a maximum amount of power produced at each instant in time. This is true for every part of the generator's rotation.

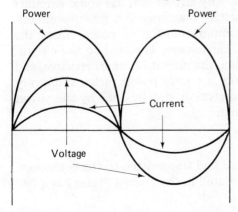

Figure 17-12. Voltage, current, and power relationships for an in-phase system.

Consider the amount of power that is created when the voltage and current are not in phase with each other. The waveforms shown in Figure 17-13 will illustrate this point. Electric power is still the product of voltage and current. The out-of-phase relationship between voltage and current produces some negative power values and some positive power values. The total amount of power that is available to perform work is less than when the voltage and current are in phase. The negative power values are subtracted from the positive values. Greater power production is required to perform an equal amount of work.

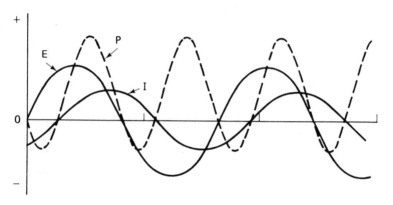

Figure 17-13. Voltage, current, and power relationships for an inductive load. Note the loss of power.

The out-of-phase relationship is usually due to quantities of inductance at the load. Typical large inductors are the electric motors used in industry. The relationship between actual power consumed and the amount of power apparently produced is called the power factor of the circuit. The ideal power factor is 1, or unity. The circuit has to be modified to correct for a high power factor. This is done by adding capacitance to the circuit. Any circuit that has equal amounts of inductive reactance and capacitive reactance is a resistive circuit. A resistive circuit permits maximum transfer of power from the source to the load. If sufficient capacitance is added to the circuit, the inductive reactance is canceled, the circuit becomes resistive, and this balance is achieved. The result is a unity power factor and a very efficient system. Power factor correction may be done by the power company or the industrial consumer.

SUMMARY

Electric power has to be created and transmitted to the consumer. Almost all the electric power generated in the United States has a frequency of 60 hertz. It is alternating-current power. The machine that produces this energy is called an alternator. The alternator creates a three-phase power at a 13,800-volt level. This power is sent from the generating plant to the consumer by means of wires. The system uses a series of transformers to transmit very high voltages at low current values.

The reason for the high-voltage and low-current ratio is found when one thinks about power losses due to the resistance of the wires. The power formula has one variation called an I^2R relationship. Power losses increase by the square of the current times the resistance. Use of the lowest possible current value minimizes these copper wire losses.

Most alternators have their three sets of coils connected in a delta or wye configuration. In a practical three-phase alternator, the currents in three of the six wires cancel each other. It is possible to eliminate these three wires when connecting the alternator to its load.

Distribution systems for electric power raise the voltage to 345 kilovolts. It is then transmitted to a series of substations. Each substation steps the voltage down to a lower value. It also branches at each substation. The end result is the 120/240 volt level found in small commercial operations and in the home.

The service to the home or small commercial establishment is 240 volts. This is distributed in the building at a distribution panel. Both 120- and 240-volt service are available. Which is selected depends upon the requirements of the load. The wiring in the panel box provides this selection. Circuit protection devices are also found in the distribution panel box.

When the power is distributed to consumers at this level, it is transmitted as single-phase power. One advantage of three-phase systems is reduction in the size of motors, alternators, and transformers from comparable single-phase components. There is another advantage. It is the ability to have 1.73 times the single-phase power when connecting to the phases of the three-phase system. This produces more power in the same wires than the single-phase system is able to produce.

A final consideration in transmission and consumption of electric power has to do with the power factor. A system that has purely resistive loads is more efficient than a system with either inductive or capacitive reactance. This is because of the relationship of voltage and current. When voltage and current are in phase, they create the highest level of power. Reactance in the system produces an out-of-phase relationship. This results in less available power at the load for an equal amount of power at the source. Transformers and motors are inductors. Their connection to the system creates a phase shift owing to inductive reactance. The addition of equal amounts of capacitive reactance components will counteract this effect. The end result is a resistive load and a highly efficient system.

QUESTIONS

17-1. Commercial power production in the United States is

    a. 50 Hz dc

    b. 60 Hz dc

    c. 25 Hz ac

    d. 60 Hz ac

17-2. Commercial power production in the United States is

    a. single phase

    b. two phase

    c. three phase

    d. none of the above

17-3. High-voltage generation and transmission is used in the United States because

    a. there are no line losses

    b. current requirements are high

    c. minimum line loss occurs

    d. power requirements are low

17-4. Wires used to carry high power are ____ compared to low power wiring.

    a. very large

    b. large

    c. medium

    d. small

17-5. Home electric service uses ____ volt systems.

    a. 60 and 120

    b. 120 and 180

    c. 120 and 240

    d. 120 and 480

17-6. Home wiring is

    a. single phase

    b. two phase

    c. three phase

    d. multiple phase

17-7. Large industries use ____ power.

    a. single phase

    b. two phase

    c. three phase

    d. none of the above

17-8. The most efficient distribution system has a ____ power factor relationship

    a. low

    b. medium

    c. high

    d. independent

17-9. Resistive loads produce a ____ power factor.

    a. low

    b. medium

    c. high

    d. independent

17-10. Electric power is the

    a. product of resistance and voltage

    b. sum of voltage and current

    c. difference between voltage and current

    d. product of voltage and current

18

Direct-Current Power Sources

The production of electric power in the United States is almost 100% alternating current. There are some instances when a direct current is required. This is particularly true in some welding applications, devices that require vacuum tubes or transistors, and the automobile. There are two ways of producing dc voltages. One of these is by the use of a dc generator. The second way uses a system called rectification to change an ac voltage into the required dc voltages.

MOTOR GENERATORS

One of the more common ways of producing a dc voltage is by the use of a dc generator. The generator is a device that is able to convert mechanical energy into electrical energy. This is done by rotating a wire through a magnetic field. A generator may produce either an ac or dc voltage. The difference is in the method of removing the EMF from the armature. A dc generator uses a segmented commutator. Its purpose is to maintain the correct polarity on the two wires that run from the generator to the load.

Generators must be driven by some outside force. The outside force provides the required rotation. The speed of rotation has a direct effect on the amplitude of the generated voltage. The voltage produced increases as the speed of armature rotation increases. The various outside forces in use to drive a generator include gasoline engines and water wheels. The most common system in use is called a motor-generator set. This type of system is shown in Figure 18-1. A

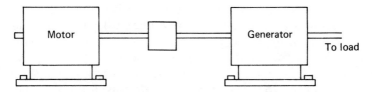

Figure 18-1. A motor-generator system for producing DC power.

motor, most often gasoline or diesel fuel powered, is operated at a constant speed. The output of the motor is a drive shaft. The drive shaft is connected to a dc generator. The constant speed of the motor maintains a constant dc output voltage.

The physical size of this combination will vary. This is dependent upon the power requirements of the load that is connected to the generator. Large power units are often mounted on a truck bed or on a trailer. Smaller units, such as those found on the automobile, are driven by a belt connected to the engine drive shaft pulley. These smaller units are mounted on brackets connected to the automobile engine.

All motor-generator sets require some maintenance. This includes lubrication of bearings and replacement of commutator brushes. There are also some problems relating to physical size and efficiency of these units when they are used in permanent installations. A much better system for use in places where ac power is available is the system using rectifiers.

RECTIFICATION

A rectifier is a device that allows current flow through it in only one direction. This device is used in power supplies designed to convert an ac voltage into a dc voltage. The block diagram of this type of system is shown in Figure 18-2. In most cases the ac source is the 60-

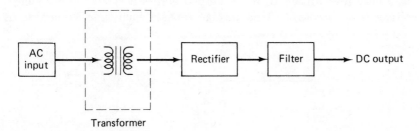

Figure 18-2. Block diagram arrangement for a power supply. This supply converts AC power into DC power.

hertz power line. The source voltage often is connected to the primary of a power transformer. The secondary of this transformer provides the required operating voltage levels. At this point the voltage is still in ac form. The rectifier changes the ac into a pulsating dc form. Electronic filtering networks smooth out any variations in the amplitude of the wave. This produces a pure dc waveform. On an oscilloscope this pure dc wave is a straight line.

There are several types of rectifying systems in use. Each is discussed in this section of the book. Assume that the power source is connected in the description of how these systems function. Filters and loads will be covered after the material relating to how rectifier systems work is explained.

Half-Wave Rectification

The simplest system of rectification is the half-wave system. It uses the device called a rectifier to allow current to flow in only one direction in the circuit. This process is called *rectification*. The half-wave rectifier system cuts off current flow during one half of the ac wave cycle. It permits current flow during the other half of the ac wave cycle. A half-wave rectifier circuit is shown in Figure 18-3. The resistor included in the circuit is shown in order to represent a load on the circuit. The ac source changes its polarity 60 times a second when the circuit is connected to power company wires. Assume that we have the ability to stop time and electricity. This new ability will aid in the explanation of how the circuit works.

At one moment in time the positive lead from the source is connected to the anode of the rectifier. At this same time the negative lead from the source is connected to the lower end of the load. In this situation the diode is forward biased. Its anode has a positive charge, and its cathode, through the load, has a negative charge. Current will flow from negative, through the load and diode, and go to the positive terminal of the source. The diode, being forward biased, has a very low voltage drop developed across it. The load has a large voltage drop across it. This voltage repeats each time the anode of

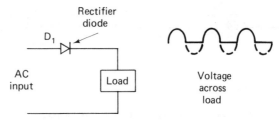

Figure 18-3.   A half-wave rectifier circuit.

the diode is positive with respect to the cathode. An output waveform observed across the load looks like the pulsating dc wave shown.
When the anode of the diode is negative with respect to its cathode,
the diode is reverse biased. Almost all the voltage in the circuit develops across the diode. There is little left to drop across the load.
During this half of the input wave cycle, the voltage observed across
the load is almost zero, and no work is performed at the load.

The polarity of the output voltage will depend upon the manner
in which the diode is wired into the circuit. When the diode is connected as shown, its cathode is connected to the positive end of the
load. The reason for this goes back to the assignment of polarities
across each component in a closed loop when analyzing voltage drops
with Kirchhoff's law. The efficiency of the half-wave rectifier circuit
is 50% of the duty cycle. There are other rectifier systems with higher
efficiencies. These are called full-wave rectifiers.

Full-Wave Rectification

The two full-wave rectifier systems are called full-wave and full-wave
bridge rectifiers. The reasons for each name will be obvious as each is
explained. The advantage of the full-wave system is that its efficiency
is double that of the half-wave system. The critical idea behind this
type of system is to rectify each half-wave of the ac cycle and to
maintain a one polarity output voltage. The ac wave constantly
changes its polarity. If one were to build a circuit using two diodes
and a load resistance, the odds are that the circuit would not work.
The only way this type of circuit works is when all the current in the
load flows in the same direction. This is accomplished with the circuit shown in Figure 18-4.

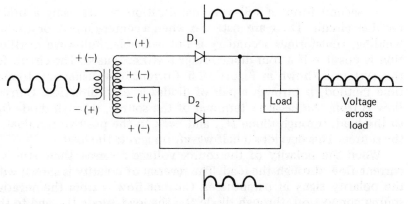

Figure 18-4. A full-wave center-tapped rectifier circuit uses both halves of the AC
alternation. The resulting waveform for this type of circuit is also shown.

In this circuit a transformer with two secondary windings is required. Often the two windings are connected together as shown in the diagram. These are called center-tapped transformer windings. Using the privilege of stopping time again, let us examine the details of the operation of this circuit. Time is stopped when the primary winding has the polarities shown. Transformer action produces a 180-degree phase shift between primary and secondary windings. The polarities for each of the two secondary windings are also shown. When these polarities are established one of the diodes, D_1, is reverse biased. The other diode, D_2, is forward biased. Current flow is from the negative lead of the transformer secondary, through the load and D_2, and back to the positive terminal of the transformer secondary. A waveform as shown below D_2 is produced across the load as this diode conducts.

Moving time ahead a little bit will set the conditions for a complete reversal of all the polarities in the transformer. These are shown in parentheses on the diagram. Under these conditions, diode D_1 is forward biased and diode D_2 is reverse biased. Current flow is now from the negative terminal of the transformer, through the load and D_1, and then back to the positive terminal of the transformer secondary. A waveform as shown above D_1 is produced across the load during this half of the input wave cycle.

Observe the phase difference between the waves developed as each diode conducts. All the current flows in the same direction through the load. As a result, a double waveform occurs across the load. This waveform is the result of combining each diode's conduction through the common load. The result is rectification during the full cycle of the input wave. This is one form of full-wave rectification.

A second form of full-wave rectification occurs using a bridge rectifier circuit. There are instances when a center-tapped, or double-winding, transformer secondary is not available. Full-wave rectification is possible if a four-diode bridge rectifier is used. The circuit for this system is shown in Figure 18-5. Current flow, using our stopped time method, is through a pair of diodes in each direction. Current flow is from the negative terminal of the source, through diode D_2, to the load, through diode D_4, and back to the positive terminal of the source. This develops a half waveform across the load.

When the polarity of the source voltage reverses, there still is a current flow through the load. The reversal of polarity is shown with the polarity signs in parentheses. Current flow is from the negative source connection, through diode D_1, the load, diode D_3, and to the positive source connection. The resulting waveforms developed across

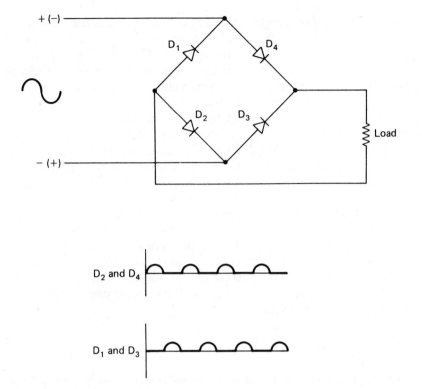

Figure 18-5. A full-wave bridge rectifier circuit. This circuit does not require a center-tapped transformer.

the load are shown below the circuit diagram. Current flow through the load occurs during both the positive and negative alternation of the source. The result is a combination of both waves across the load. This produces a full-wave shape across the load.

These two forms of full-wave rectification are both found in electrical circuits. One question remains to be answered when discussing rectifier circuits. This is related to the relationship between input voltages and output voltages. A general statement about this relationship is that the output voltage is equal to the peak value of the input wave.

Let's review terminology related to ac waves. One has two methods of describing the same voltage value. One method compares the ac voltage with a comparable dc voltage. This is called the rms value. It is discussed earlier in this book. The second method is to measure the amplitude of the peaks of the ac wave. This provides a peak-to-peak voltage value. Conversion from rms to peak-to-peak voltages is accomplished by multiplying the rms value by 2.828. The result is the peak-to-peak value for the voltage.

A rectifier system allows current flow in one direction through the circuit. An ac voltage is applied to the circuit from the source. Half of this voltage develops across the load. Therefore, half of the rms value is developed across the load. It is more convenient to convert the rms voltage into a peak-to-peak voltage at this point. If a rms value of 100 volts is applied to the circuit, the peak-to-peak value is 283 volts. Half of this voltage appears across the load due to rectifier action. Half of the 283 volts is 141.5 volts. This is the peak output voltage that is obtained from this circuit. Remember, this is a pulsating dc voltage. It rises to a maximum value and falls to zero during each half of the alternation. Further circuit action is required to make this voltage a constant-value dc voltage. The circuit used for this purpose is called a filter circuit.

FILTER CIRCUITS

The waveforms that developed across the load resistance are not a pure dc form. A pure dc waveform is illustrated in Figure 18-6. At time zero the voltage has zero amplitude. When the circuit is turned on, the voltage rises to its full value. It remains at this level until turned off. When off, the voltage in the circuit drops back to the zero level.

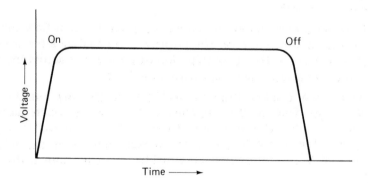

Figure 18-6.   The voltage values that occur when the power is applied in a DC circuit.

The output voltage of a rectifier system does not resemble this waveform. It looks more like that shown in Figure 18-7. These waves

Pulsating DC voltages

Half-wave   Full wave

Figure 18-7. Pulsating DC voltage waveforms. Half-wave and full-wave values are shown.

are called pulsating dc voltages. They have to be modified in order to act more like a pure dc voltage waveform. The pulses have a very definite frequency. Half-wave pulse rate is 60 hertz, and the full-wave rate is 120 hertz. An electric filter system is used to smooth the pulsating waves. The filter is made up of capacitors and inductors. Typical connection for the capacitors is to wire them in parallel with either the source, the load, or both. Inductors are usually series connected between the source and the load. Circuit diagrams for three typical filter circuits are shown in Figure 18-8. The names for each of these filter systems are fairly self-descriptive. The circuit on the left is called a capacitor input filter. The center circuit shows a choke, or inductor, input. The right hand circuit illustrates a pi (for the Greek letter π) section filter.

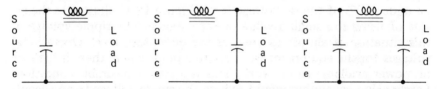

Figure 18-8. Diagrams for electronic filter systems used with power supplies.

Capacitors in a filter system attempt to smooth any voltage variations in the circuit. The action of a capacitor is shown in Figure 18-9. The output of a half-wave rectifier is shown in the upper part of the illustration. The addition of a capacitor in parallel with the load produces a waveform that looks like that shown in the bottom of the figure. Capacitor action is like a spring. Rising voltage is absorbed by the capacitor. During a fall in voltage, the capacitor gives up its charge to the circuit. This action results in a gradual decline in the circuit voltage. The time constant of the capacitor is long. It is not able to completely discharge. As a result, a pulsating voltage is developed. This voltage does not drop to zero, but varies at a level that is near the operating voltage of the set. This type of variation of the dc oper-

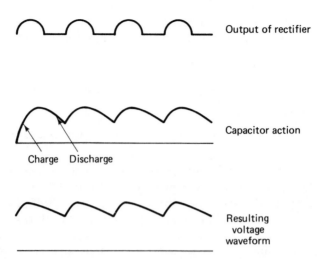

Figure 18-9. Capacitor action in a filter circuit fills in the voids left when the diode is not conducting.

ating voltage is called a *ripple* voltage. Its frequency is dependent upon the frequency of the rectifier system.

The amount of ripple voltage is dependent upon the size of the capacitors used in the filter system. The ripple voltage created by capacitor action is apt to adversely affect the operation of a piece of equipment. In most situations the amount of ripple voltage should be under 10% of the dc voltage level. Figure 18-10 illustrates the result of using too small a value of capacitance. The ripple voltage in this situation is almost as great as the dc voltage level. Operational voltages for the equipment will vary to a point where there is little or no power available to do work. This is a very undesirable condition. Larger-value capacitors would reduce the ripple voltage to an acceptable level.

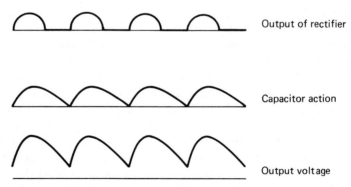

Figure 18-10. Capacitor action with a capacitor that is too small for the circuit results in a large ripple voltage.

Inductors react to changes in circuit current in the same manner as the capacitor reacts to voltage changes. The magnetic field created in an inductor will tend to maintain the circuit current just as a flywheel maintains rotary motion in a machine. Current variations occur owing to the changing voltage levels produced by the rectifier. The current flow in the circuit is established at a level determined by Ohm's law. This level is maintained by the flywheel effect of the inductor.

The combination of capacitance and inductance produces a filter effect on the circuit. Any variation on the amplitude of voltage or current is not allowed to pass through the filter. The filter is actually a low pass frequency filter. It only allows a very low frequency (and dc has a frequency of zero) to pass through its components. All other frequencies are filtered out and do not pass on to the load.

REGULATION

One factor to consider that is directly related to dc power supplies is the regulation capability of the supply. This refers to the ability of the power supply to maintain a constant output level under varying load conditions. Each dc power supply has some limitations. These limitations are determined by the power rating of the components used to construct the supply.

In a practical dc power supply, a decrease in output voltage occurs as the load is increased. An increase in the load is inversely proportional to the ohmic value of the load. A decrease in load resistance produces a higher level of current flow in the circuit. The practical power source does not have unlimited current capabilities. As a result there is a gradual decrease in the voltage of the power source as the load increases. At zero resistance, or short-circuit conditions, the output of the source drops to zero volts.

There are a variety of circuits and devices in use that act as voltage regulators in circuits. Current technology has developed some of these as microelectronic circuits. All the components required for electronic regulation are packaged in one small plastic or metal housing. These devices have three connections to the electric circuit. One such device is shown in Figure 18-11 along with a typical circuit diagram. These voltage regulators are available in a range of values from about 6 to 24 volts. The current capacities will range up to 5 amperes. Regulators capable of handling other voltages and higher currents are not available in this type of package. They can, however, be constructed from individual components. The purpose of the electronic regulator is to maintain a constant output voltage over a range of

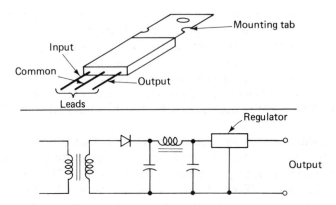

Figure 18-11. A solid-state voltage regulator and its electronic circuit schematic diagram.

load resistances. This is possible with most regulators until values are exceeded. In many cases the regulator turns off the output current when overcurrent conditions exist.

SUMMARY

Some electrical devices require a dc voltage source. Almost all the power requirements in the United States are met with an ac source. Consequently, there has to be a way of producing dc voltages. This may be accomplished in one of two ways. One way is to use a motor to drive a dc generator. The output voltage of the generator has to be maintained by controlling the speed of the generator. This works well in portable or remote applications.

The heavy power requirements of industry require another means of producing dc voltages. This is accomplished by using a rectifier system. This system changes the ac voltage into a pulsating dc voltage. Three circuits in use today are the half-wave, full-wave, and full-wave bridge rectifiers. Each has the ability to produce the required pulsating dc from an ac source.

A semiconductor device called a diode is used for rectification. The diode conducts during the half-cycle when it is forward biased. The diode is connected to the load in order to develop power in the load during its conduction cycle. The result is a half-wave power source.

Full-wave rectification is possible using either two or four diodes in a circuit. Two diodes are used with a center-tapped transformer. Four diodes are used in a bridge circuit when no center-tapped transformer is available.

The output voltage from the rectifier system is a pulsating dc. The frequency of the pulsation is 60 hertz for the half-wave system and 120 hertz for the full-wave system. The desired dc voltage without pulsation is obtained by adding a filter network to the output of the rectifier. The filter components are capacitors and inductors. The capacitors oppose any change in circuit voltage. Inductance opposes changes in circuit current. The combination of capacitance and inductance provides a filter circuit. This particular filter allows only very low frequencies to pass through it. All other frequencies are stopped by the filter action.

Another point to consider is the regulation of the output voltage. Each power source has limitations. Under normal operation circuit voltage tends to drop as the resistance of the load decreases. Voltage could drop to close to zero under short-circuit conditions.

Special voltage-regulation devices are available. These regulators are designed for low-voltage applications. Their current range is from less than 1 ampere to about 5 amperes. Higher voltage and current regulators are built from discrete components.

QUESTIONS

18-1. A generator

    a. converts electrical energy into rotary energy

    b. converts ac into dc

    c. produces small amounts of power

    d. converts magnetic energy into electrical energy

18-2. The process of changing ac into dc is called

    a. switching

    b. rectification

    c. transferring

    d. time magnification

18-3. The process of changing ac into dc produces

    a. smooth dc

    b. one-way ac

    c. pulsating dc

    d. negative power

18-4. The system that uses two diodes is called a ____ system.

    a. half-wave

    b. full-wave

    c. full-wave bridge

    d. bridge

18-5. The system that uses four diodes is called a ____ system.

    a. half-wave

    b. full-wave

    c. full-wave bridge

    d. bridge

18-6. The voltage develops across the load when the diodes are

    a. forward biased

    b. reverse biased

    c. connected to the filter

    d. independent of the source

18-7. The efficiency of the one-diode system is ____ of the full duty cycle of the sine wave.

    a. 25%

    b. 50%

    c. 75%

    d. 100%

18-8. The efficiency of the four-diode system is ____ of the full duty cycle of the sine wave.

    a. 25%

    b. 50%

    c. 75%

    d. 100%

18-9. The function of a filter network in a power supply is to

    a. supply a varying voltage

    b. supply a smooth voltage

    c. remove voltage and current variations

    d. provide an ac component to the load

18-10. In a practical dc power supply, ____ output voltage occurs as the load is increased.

    a. an increase in

    b. no change in

    c. a varying level of

    d. a decrease in

19

Schematic Diagrams

The ability to communicate is essential in any society. This is particularly true in a highly technical society. Persons employed in electrically related occupations have a very definite need to have some common basis for effective communication. One factor in any communication system is a language that is understood by all. Many languages are spoken in our world. There is a need for a base upon which electrical circuit communications are founded. This base is the electric schematic diagram.

The schematic diagram has two features that allow it to be used for communication when a written or spoken language is not common. These two features are the use of a standard set of graphic symbols and an almost universal layout format for the diagram. This chapter is devoted to the explanation of graphic symbols and to the manner in which a schematic diagram is read.

GRAPHIC SYMBOLS

The common language in electrical work is the set of graphic symbols. These symbols have been developed over the years by persons in the industry. Each symbol is related to a specific component. Components often are produced in a variety of sizes and shapes. The purpose of the schematic symbol is to provide a representation of the function of the device. A parts list that relates schematic symbol identification to an exact make or model number for each component is required in order to discover size and shape of each component.

Standardization of graphic symbols is accomplished by the electrical industry. A set of graphic standards has been developed by the National Electrical Manufacturers Association (NEMA). A second set of symbols has also been developed by the National Machine Tool Builders Association specifically for machine tools. The second set of standards has evolved into a set called JIC standards. This represents the work of a Joint Industry Council. Both of the graphic standard sets are similar in most respects. Each has its own minor differences, also. It is not the purpose of this book to promote either set of standard symbols. Due to the similarities of each set, only one set will be used. This set is the one presented by the JIC organization. The complete set of graphic symbols is shown in Figure 19-1. Selected symbols shown in this illustration are explained in this section.

The first portion of the symbols shows a variety of switches. A small circle is used to represent a contact on the switch. A solid line that is drawn to one of these contacts indicates a wire or a part of a contact assembly other than the actual contact. Refer to the disconnect switch in the upper left corner of the symbol set. Switches and other moving part assemblies are always shown on a schematic in their "at rest" position. This is illustrated by the connector line between the contacts being open in this drawing. The dashed line is used to indicate that the three switches are mechanically connected, but not electrically connected. In other words, when the switch is turned to its "on" position each set of contacts is connected to its corresponding mate. Three separate circuits are turned on by the action of this specific switch.

Another often seen symbol is shown in the lower right corner of this same page. The standard shown here illustrates the method of drawing wiring that crosses on the paper but is not connected. Also shown is wiring that connects and/or crosses and connects.

On the top of the second page of symbols is a set used to show relay contacts. When the contacts are normally open, or separated, their lines do not touch. When the contacts are normally closed a diagonal line is drawn across them. Figures 19-2 through 19-5 show typical components and their graphic symbols.

SCHEMATIC DIAGRAMS

Most electrical schematic diagrams are presented in order to show how the circuit functions. There is very little relationship between the actual physical layout of the unit and its schematic diagram. A second role that the schematic diagram plays is that it shows the sequence of events as they occur during the operation of the device.

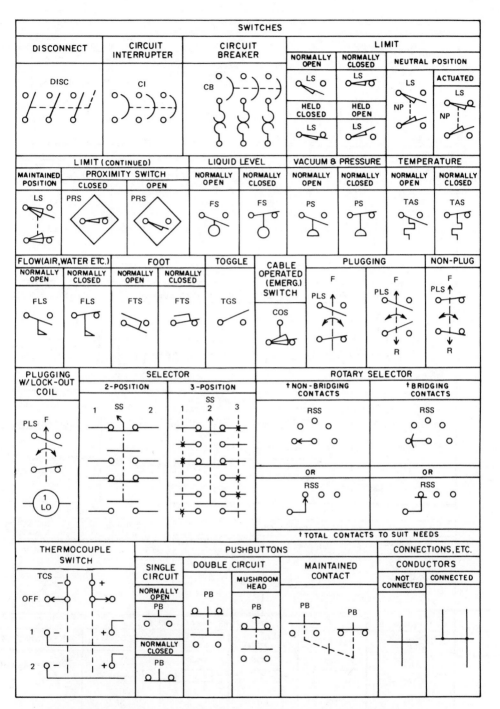

Figure 19-1. Electrical schematic symbols used for machine tool wiring. These are called JIC symbols.

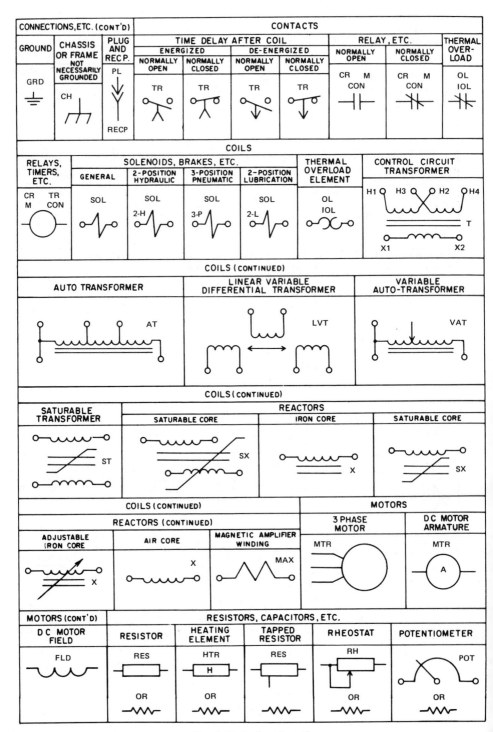

Figure 19-1 *(continued)*

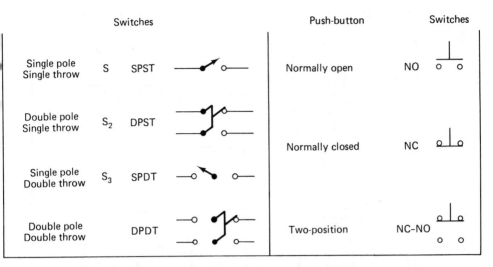

Figure 19-2. Schematic symbols for switches (JIC).

Very often the schematic diagram is divided into sections. Usually the upper portion of the diagram shows the power control devices and the load. The lower portion of the diagram shows the controlling circuitry. This is illustrated in Figure 19-6. Connections and components are identified on the schematic. The set of standard symbols provides this information. When more than one component of the same kind is used, a number is assigned to each component. This is done in order to identify individual components. The number may be placed on either side of the letters used to represent the device.

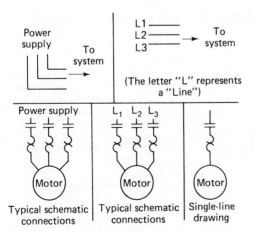

Figure 19-3. Schematic symbols for wiring systems (JIC).

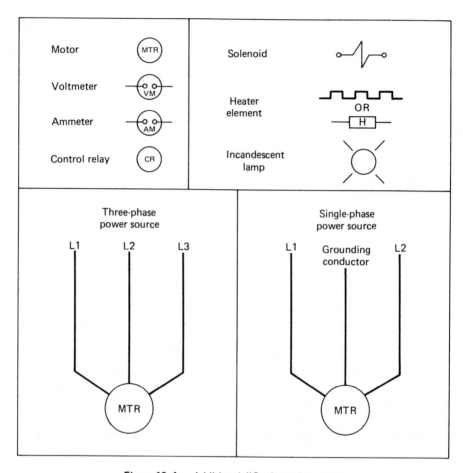

Figure 19-4. Additional JIC schematic symbols.

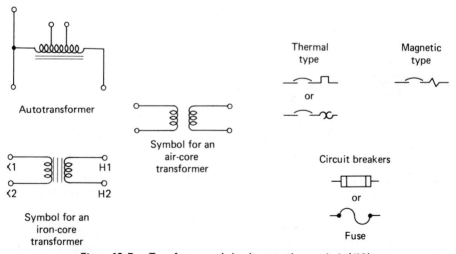

Figure 19-5. Transformer and circuit protection symbols (JIC).

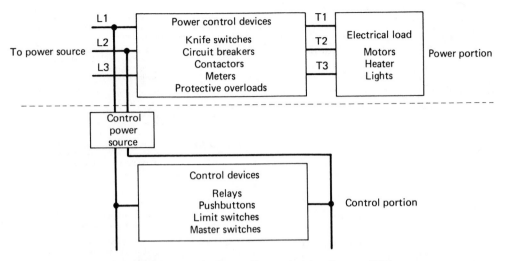

Figure 19-6.   Layout for machine tool wiring diagrams (JIC).

For example, a group of contactors could be identified as 1C, 2C, and 3C. If the contactor has more than one set of contacts, each set is assigned the same number. This may appear anywhere on the schematic diagram. Wherever it appears gives reference to the same unit.

Before examining a typical schematic diagram, a review and explanation of component function and schematic symbols is in order. Figure 19-7 offers a reasonably complete explanation of the function of each graphic symbol. Use this chart and the chart showing graphics for components in order to assist you in learning how to read the schematic diagrams. There are several electrical schematics shown in the book in the remaining chapter. This material will help you in understanding them.

The schematic diagram shown in Figure 19-8 will illustrate how this system works. The source of power is shown on the upper left side of the schematic. The three lines marked as L_1, L_2, and L_3 represent a three-phase source. Most schematics have a flow path in the same order as one reads. The action moves from the left of the drawing to the right of the drawing. This drawing happens to be fairly standard for control of a machine tool. A second flow is created. This flow path is for the control devices. The action path for control circuits is a vertical path.

Examine each path of the schematic diagram given in Figure 19-8. Start with the major current path. This runs from the source to the load. A single-phase transformer primary is connected across L_1 and L_2. This supplies the correct power for the control circuits. Moving on toward the load, the next symbol shown is a circuit breaker. One circuit breaker is indicated by the dashed lines. Next in line is a set of contacts. These contacts are identified by the letter M. Each of the three contacts has the same letter as an identifier. This indicates that

Symbol	Part	Component	Function
(or)	Air-insulated contacts	Circuit breaker CB	Opens or closes circuit by nonautomatic means and automatically opens circuit at predetermined overload current.
	Oil-insulated contacts		
	Coil	Contactor C	Makes the breaks power circuit to the load when coil is energized or deenergized.
(Same as control relay)	Contacts		
	Coil	Control relay CR	Energizes or deenergizes electrically operated devices when coil is energized or deenergized.
	Normally open (NO) contacts		
	Normally closed (NC) contacts		
	Winding and primary	Current transformer CT	Induces low current in secondary wound around primary.
	Float and NO contact	Float switch FS	Makes or breaks circuit when actuated by float or other liquid-level detector.
	Float and NC contact		
	NO contact	Foot switch FTS	Makes or breaks circuit when manually actuated.
	NC contact		
	Entire fuse	Fuse FU	Breaks circuit when predetermined current melts conducting element.
	Contacts	Knife switch KS	Makes or breaks circuit when manually engaged or disengaged.
	NO contact	Limit switch LS	Makes or breaks control circuit when mechanically actuated by motion or part of powered machine.
	NO contact held closed		
	NC contact		
	NC contact held open		
(See Fig. 2)	Coil, overload, relay, and contacts	Motor starter M	Makes or breaks power circuit to motor when coil is energized or deenergized.
	Magnetic coil	Overload relay OL	Disconnects motor-starter coil at predetermined overcurrent.
	Thermal element		
OL	Contacts		
	Lamp	Pilot light	Visually indicates presence of voltage in a circuit. (Color is indicated by initial.)
	Lamp and switch	Push-to-test pilot light	Combines pilot light with self-testing circuit and switch.
	NO contact	Pressure switch PS	Makes or breaks circuit when pressure or vacuum bellows actuates switch contacts.
	NC contact		

Symbol	Part	Component	Function
	NO contact	Push-button switch PB	Makes or breaks circuit when plunger is manually depressed or released.
	NC contact		
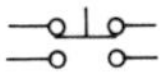	Combined NO and NC contacts		
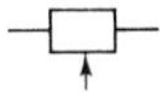	Terminals and slider	Rheostat RH	Reduces voltage in a circuit by passing current through resistance winding and tapping the reduced voltage from a sliding contact.
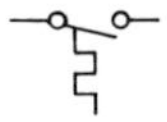	NO contact	Temperature switch TS	Makes or breaks circuit when a predetermined temperature is attained.
	NC contact		
	Tie point	Terminal T	Provides convenient connection point.
	NO timed-closing (TC) contact	Timing relay TR	On delay retards relay-contact action for predetermined time interval after coil is deenergized
	NC timed-opening (TO) contact		
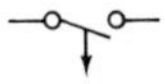	NO, TO contact		Off delay retards relay-contact action for predetermined time interval after coil is deenergized
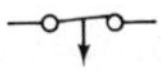	NC, TC contact		
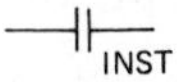	NO instantaneous contact		
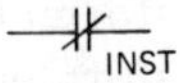	NC instantaneous contact		
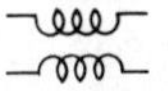	Windings	Voltage transformer	Reduced or increases circuit voltage when current in primary winding induces voltage in secondary winding.

Figure 19-7. Explanation of symbols, parts, and their respective functions.

the three contacts are operated by a common device. This device, which is in the control section, also uses the letter *M* in order to relate it to the contact assembly. A set of overload relays appears next on the schematic. Reference to the symbol chart shows that these are the coils for overload relays. These coils are thermally controlled. In other words, they open at some predetermined temperature. They are designed to shut off power to the motor if the assembly gets too hot owing to excess current flow through it. The final symbol on this part of the schematic shows a three-phase motor.

This part of the circuit is similar in operation to a series circuit. Each line from the source to the motor has to be operational before

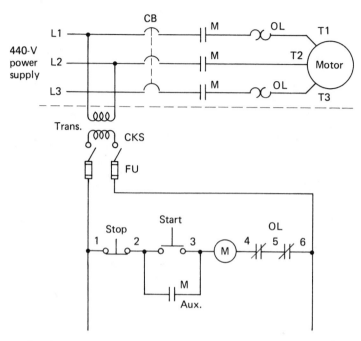

Figure 19–8.   A sample wiring diagram as explained in the text.

the motor will run. If any one of the three lines is open, the motor is not going to function correctly. None of the action will occur until the controller circuit sets it up for action.

The control circuit obtains its power form a control transformer. The letters *CKS* refer to the control designation for this transformer. Control wiring runs vertically on each side of the schematic for this section. In some cases there are many sequential control functions. Common procedure is to show these in order of their sequence. The circuit closest to the source is the first in the sequence. Those farther away from the source or the schematic occur in the order of the distance away from the source.

Each line of the secondary of the control transformer is connected through a switch and a fuse to its respective side of the control line. On this simplified diagram there is only one control function. The control circuit is in series across the lines of the control transformer. Each device has to be activated in order to close the contacts on the line to the motor.

Moving from left to right one finds a *stop* switch. This switch is normally closed. It is on in the "rest" position. The next item on the line shows two devices connected in parallel. These symbols are for a magnetic starter. The start switch engages an electromagnet. The electromagnet holds a set of contacts in the closed position. Should a power failure occur, the contacts in the electromagnet drop out.

They remain off until the start switch is reactivated. This is a safety feature. It protects both the operator and the machine in case of a shutdown.

The next device is the coil for the line contactors. This is another electromagnet. When it is energized, the contacts *M* on the upper part of the schematic are connected. Power then is applied to the motor. The remaining two sets of contacts represent parts of the overload relays. These open if the temperature of the overload relay coil exceeds its setting. All the components on this line are dependent upon each other. The only exception is the magnetic circuit around the start switch. Since this switch is normally open, the current path for this line of the controller is through the coil and contacts. If any one part of this current path should open, the contacts on the line (*M*) also open. This shuts down the motor by breaking the circuit current path to the motor. This is an example of one rather basic motor control circuit.

The schematic diagram used as an example in this section of the chapter uses a form agreed upon by the Joint Industry Committee. It is specifically designed to show machine tool operation. Included as an essential component in this diagram is a sequence of operation for the machine. There are other types of schematic diagrams in use by the electrical industry. Some of these follow the same general layout format used for machine tools. The exception comes when one tries to determine a sequence of events.

Many machines that use electricity are not programmed for sequential events. This is particularly true in the automobile. The operator of the vehicle has several options available. These include operation of headlights, radio, windshield wipers, and climate control systems. The schematic diagram for the electrical system of the automobile shows the capability of selecting these options. One thing that a person reading a schematic diagram has to know is how the equipment operates. Once this is understood, a review of the electrical schematic will be meaningful. Circuits using these types of schematic drawings are also found in the remaining chapter of this book.

SUMMARY

One form of communication that is fairly standard in the electrical industry is the schematic diagram. This diagram differs from a wiring diagram. The schematic diagram shows the scheme of the operation rather than the manner in which specific parts are wired to each other.

A set of graphic symbols is used by the electrical industry. The symbols represent specific components used in electrical circuits. Each component's size, shape, or capability is not shown on the schematic diagram. This information is found on a parts list for the device.

Machine tool builders have developed a set of drafting standards. This set is called the JIC standards set. Included in the standards are graphic symbols and a method of layout for the schematic diagram. Each schematic diagram has some plan for showing the operational path of the device. The JIC standards use a horizontal path to represent a function. Vertical paths show the sequence of events. Schematic diagrams are read in a manner similar to that used when reading a book. The upper part of the diagram is used to show the source, the load, and power control devices. The balance of the diagram is used to show a sequence of events in the control portion of the machine. Each line in the control section represents one step in a sequence of events. The line at the top of the diagram is the first in the sequence.

Schematic diagrams used for other electrical devices often follow this general format. Not all devices require a specific sequence of operation. Some devices offer an operator the opportunity to select one or more of many options. It is important that a person working on electrical devices know how the device is supposed to function. This information will aid in the reading and understanding of the schematic diagram.

QUESTIONS

19-1. Graphic symbols are used to represent

    a. resistors

    b. capacitors

    c. inductors

    d. all of the above

19-2 Most schematic diagrams read from

    a. right to left

    b. top to bottom

    c. bottom to top

    d. answers a and c above

    e. answers b and c above

19-3. Most schematic diagrams show

    a. a sequence of events

    b. a path for current flow

    c. symbols instead of illustrations for parts

    d. all of the above

19-4. Wires that cross and connect are shown on the schematic diagram by

    a. use of a dot when the lines cross and connect

    b. use of a jump-over connection

    c. a written note on the schematic

    d. use of a separate wiring diagram

19-5. Component identification on the schematic uses

    a. letters

    b. numbers

    c. letters and numbers

    d. none of the above

19-6. Power sources are usually drawn on the ＿＿ side of a schematic.

    a. lower left hand

    b. lower right hand

    c. upper left hand

    d. upper right hand

19-7. Wiring from the power source to the load is normally a ＿＿ wiring circuit.

    a. series

    b. parallel

    c. multipath

    d. isolated

19-8. The automobile wiring diagram is an example of a ＿＿ wiring system.

    a. series

    b. parallel

    c. isolated

    d. secondary

19-9. The JIC schematic diagram uses the vertical path to show

    a. motor action

    b. function

    c. starting action

    d. event sequence

19-10. When reading a schematic diagram, one must know

    a. schematic symbols

    b. device function

    c. electrical theory

    d. answers a and b above

    e. all of the above

20

Control of Electric Power

Most of the people living in an industrialized society have their lives affected by electricity on an almost continuous basis. This influence has become so accepted that we feel that electric power is always with us. Little, if any, thought is given to the how or the why of electric power. When a switch is turned on, we have come to expect certain results. Our only time of concern is when a switch is turned on and the unexpected happens. Then our thoughts will focus on the how and the why as we attempt to return the controlled device to a functional condition.

Material presented in this book is concerned with both the how and the why of an electric device. We have dealt with many concepts in an attempt to explain electrical theories. A good many examples of both how and why are given in preceding chapters.

This chapter discusses how electric power is controlled. We have learned how to use electric power to perform some function. The function is often classified as work. The thought behind this term is that electrical energy is converted into another energy form. The second energy form is actually the performance of a function. This function is called work. In actual practice, work is performed by the electric power that operates a machine. Work is also performed by the electrical energy required to operate a television set or any home appliance.

Whenever some work function is being performed it has to have some capability for control. In a good number of situations, automatic control systems have been developed. In other cases direct control of the electric power must be used. Electric power control is accomplished by a variety of methods. These include timing systems, wave shape systems, amplitude control systems, and some automatic systems. Also included in the control system are systems that use remote controls. Each type of system is discussed in this chapter. A block diagram of a typical control system is shown in Figure 20-1. This general format is developed into more specific systems. Diagrams and illustrations for each system are presented. In each type of system the purpose is to provide a method of controlling electric power. The capability to control electric power and to master it for use by human beings provides us with an upper hand in utilization of electric power instead of muscle power. We have also been able to increase the amount of work being performed by controlling this energy. Each of the next sections of this chapter explains how this is accomplished.

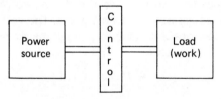

Figure 20-1. A block diagram for an electrical circuit showing the control function.

AMPLITUDE CONTROL

Power in electrical terms is the product of voltage times current. The term voltage refers to the amplitude of the electromotive force that is applied to the load. The term current refers to the amount of current that is flowing through the load. There is no current flow through the load until the source of EMF is connected to the load. It is possible to control the amount of work at the load by controlling the EMF applied to it. This may be accomplished in several different ways. Let us assume that the source EMF is a constant value in each of these examples. This assumption if based on the development of the EMF at some distant generator. Each of the control systems described is connected between the source and the load.

One method of amplitude control of EMF is to establish a voltage-dividing network. One such network is shown in Figure 20-2.

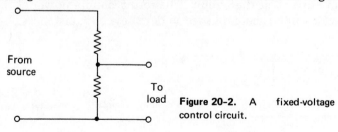

Figure 20-2. A fixed-voltage control circuit.

This network consists of two resistors connected in series. The two-resistor network is connected to the source. One side of the source is connected to the load. The junction of the two resistors is also connected to the load. The voltage developed at the junction is of a lesser amplitude than the source voltage. This is based on Kirchhoff's voltage law. The ratio of the two resistor values determines the value of the EMF that is applied to the load. The limitation of this circuit is that the only way that the load EMF can be changed is by changing the values of the resistors.

An adaptation of this circuit is shown in Figure 20-3. The two fixed-value resistors are replaced with one variable resistance. The total resistance value is connected to the source. The movable arm of the variable resistance is connected to the load. The other load connection is made to one of the resistor terminals, which is also connected to the source. The arm is able to be moved to any position on the fixed resistor. This produces the same effect as did the two resistors described earlier. The advantage of this circuit is that the voltage to the load may be varied. One does not have to change resistor values in order to change the load EMF value.

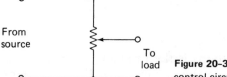

Figure 20-3. A variable-voltage control circuit.

Still another way of controlling load EMF is shown in Figure 20-4. In this circuit a variable resistance is connected as a rheostat. The rheostat is connected in series with the load. A voltage drop is developed across each component in any series circuit. This is caused by the current flow through the circuit. The amount of voltage drop is directly proportional to the comparative sizes of the resistances. The load acts as a resistance as far as the circuit is concerned. Changing the size of the variable resistance will vary the voltage drops across each of the two resistances. A large voltage drop across the load as compared to the voltage drop across the rheostat will provide an increase in current in the circuit. The result is more power at the load. More work will be accomplished under these circumstances.

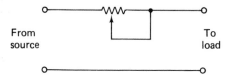

Figure 20-4. A rheostat control circuit.

A variation of the rheostat circuit is shown in Figure 20-5. A transistor is used instead of the rheostat. The emitter-collector resistance is varied by adjusting the variable control R_1. The voltage applied to the base of the transistor controls the emitter- collector resistance. This changes the voltage drop across the transistor and the voltage drop across the load. The circuit consisting of the transistor, the load, and the source is a closed loop. Kirchhoff's voltage law applies here, also. The voltage drops developed across each resistance will add to equal the source voltage. This circuit has one limitation that is not found with the other circuits. It only works on a dc source. The other circuits function with either an ac or a dc source.

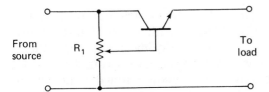

From source R_1 To load

Figure 20-5. A solid-state control circuit using a transistor.

A circuit that works only with ac is shown in Figure 20-6. It uses a variable inductor instead of a resistance. The inductor has some advantages over each of the previously described devices. It does not produce heat as a result of its opposition to current flow. Each of the other devices generates considerable amounts of heat as it offers an opposition to the current flow. This heat is wasted energy. It requires some means of dissipation to keep the device operational. In some situations, fans have to be used to circulate the air and keep the devices cool.

The inductor shown in this illustration does not create heat as it operates. The total inductance in the circuit is changed as the iron core is moved in and out of the coil. As the core is moved into the center of the coil, the inductance increases. The core directs the magnetic lines of force created by the current flow. Each magnetic line of force cuts across another turn of wire on the coil. This produces a counter EMF in the coil. The counter EMF has a polarity that is op-

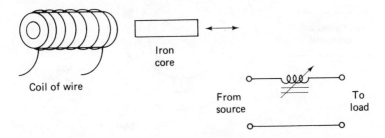

Iron core

Coil of wire

From source To load

Figure 20-6. Use of a variable inductor to control current in the load circuit.

posite to that of the EMF applied to the conductor. These two voltages tend to cancel each other. The result is a circuit that acts as though the source EMF is reduced. This reduces the EMF applied to the load without creating any undesired heat.

When the core is moved out of the coil, the inductance in the coil is reduced. Fewer lines of force are cutting across their neighboring wires in the coil. A smaller counter EMF is created under these conditions. The result is a higher value of EMF at the load. More current flows and more work is accomplished.

TIMERS

A problem that has been present in the electric control field for many years is that of time delay. Many machines found in the home and in industry require some form of time delay in their operation. Some of these machines require only a delay in starting. Others are more sophisticated. They require some predetermined sequence of events. Both of these requirements may be met by use of a timer.

Mechanical Timers

One type of timer is the mechanical timer. It uses a mechanism that drives a mechanical sequencer. The drive unit is often a small synchronous motor. In some all-mechanical units the drive element is a spring-wound motor. In any case, a drive unit is required to establish some element of time in the system.

An illustration of a basic mechanical timer is shown in Figure 20-7. This system uses an electric motor to drive a timing disc. The disc is made of a nonconducting material. On one of its surfaces is a conductive pattern. The pattern is laid out so that it corresponds to the timing sequence desired. Two sets of contacts ride on the disc. One

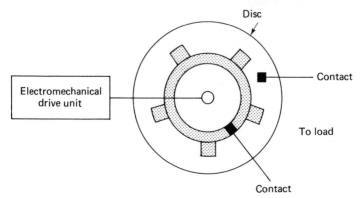

Figure 20-7. A basic mechanical timer uses a rotating disc contact plate.

of these is always in contact with the conductive pattern. The other contact acts as a switch. As the disc rotates, parts of the conductive pattern touch it. When these are in contact, the circuit is complete and the load is operating. Continued rotation of the disc separates the contact from the pattern. Under these conditions the circuit is turned off.

This type of timer is found in devices where the duty cycle is not critical. Some of these timers are used in washing machines and clothes dryers. They are dependable and relatively inexpensive to build.

Electrothermal Timers

One of the simpler types of timers is shown in Figure 20-8. This is an electrothermal timer. It uses a heating element and a bimetallic contact strip. The bimetal strip is made of two pieces of metal. Each metal has a coefficient of expansion. This term refers to the amount of expansion that occurs when the metal is heated. Two metals having different expansion rates are bonded together. When they are heated, the bonded strip will bend to form an arch. This principle is used to make a set of contacts touch each other. It may also be used to separate a set of contacts.

In this example, the contacts will "make" when the bimetal strip is heated. The contacts are connected in series with the load. A control circuit consisting of a heating element is connected to the source. When the operating power is turned on, a current flows through the heater. The resulting heat bends the bimetallic strip and closes the contacts. The contacts complete the power circuit to the load. This type of circuit is limited by the heat created in the control circuit and the position of the contacts. Another limitation is cycle time.

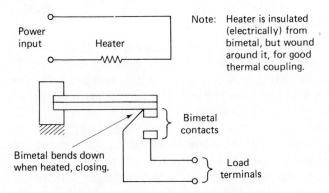

Figure 20-8.   A bimetal contact timer depends upon heat to move the contact points.

The circuit cannot be reset until the bimetal strips and heater are cool. This usually takes about the same time as the time required to heat the unit. The time span for a circuit of this type runs from a low of about 10 seconds to a maximum of a few minutes.

Electromechanical Timers

One of the most common forms of timers is shown in Figure 20-9. This is an electromechanical type. A small synchronous ac motor drives a gear train. The gear train drives, through a set of reduction gears, a cam. The cam operates a mechanical switch. The switch contacts control the power to the load.

Some electromechanical timers are designed to automatically cycle during a 24-hour period. Other types are designed to be reset after their operational cycle. The operation for a specific timer is left to the circuit designer.

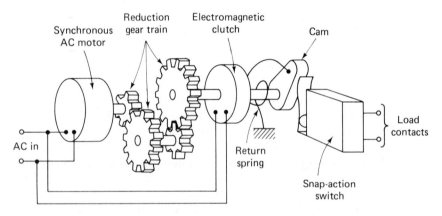

Figure 20-9. An exploded view of an electromechanical timer.

Electropneumatic Timers

A third type of timer uses both electrical and fluid power principles for its operation. This is the electropneumatic timer. This timer has a restricted air chamber. The length of the chamber is determined by the position of the time control. The time control setting adjusts for the length of time of operation of this device.

A solenoid is connected to a pneumatic, or air chamber, diaphragm. Activation of the solenoid moves the diaphragm. This creates the air pressure chamber in the timer. The pressurized air will slowly escape from the chamber. Also attached to the solenoid is a set of contact points. When the air pressure in the chamber is reduced to a low level, the contact points "make." This completes the control circuit wiring and operates the machine.

Electronic Timers

There are different types of electronic timers in use today. One of these is shown in Figure 20-10. This circuit uses the RC time-constant principle. A resistor and a capacitor are series wired across a source. When power is applied to the circuit, the capacitor charges. The length of time required to charge the capacitor is dependent upon the values of the resistor and the capacitor. The formula $T = RC$ (time, in seconds = resistance, in ohms, times capacitance, in farads) is used to determine the charge time of the circuit. One time-constant period is equal to the time required to charge the capacitor 63% of its full charge value. This threshold voltage is the value at which a circuit is activated. This circuit is called a *threshold* circuit because of its activation point.

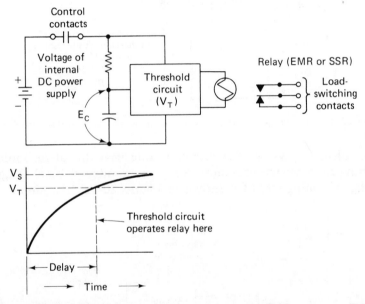

Figure 20-10. Charging of the capacitor turns on the threshold circuit. This, in turn, applies power to the relay.

The threshold circuit may use one of several devices. The circuit may activate a transistor, a SCR, or a thyristor. These controlling devices, in turn, operate a relay or other power-switching device. Circuit designers may select an electromechanical device, which is a relay, for the device that operates the load. Another choice is from the thyristor family. Here the choice is limited to the SCR in a dc circuit or a triac in an ac circuit. In any case, a threshold circuit is used to trigger the load-controlling circuit.

Various threshold circuits are also used. The simplest of these is the *RC* circuit previously described. More complex circuits are also used. Some of these use other solid-state devices for the threshold circuit. One such device is called a *unijunction transistor*. This transistor acts as a switch when a specific voltage is reached at its gate. The transistor switches on and operates the thyristor. In this way a small amount of current at the transistor can be used to control the heavier currents in the thyristor. This is illustrated in Figure 20-11.

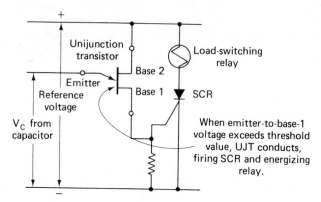

Figure 20-11.   In this circuit the voltage on the transistor turns on the gate of the SCR.

An ultimate system for circuit timing uses digital technology. This type of timer requires more complex circuitry. A block diagram of a digital timing circuit is shown in Figure 20-12. The circuit uses

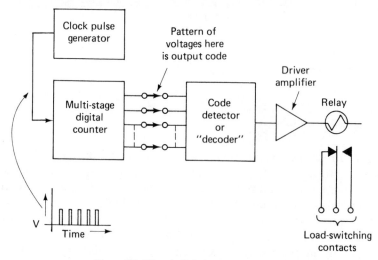

Figure 20-12.   A digital timer circuit.

timing pulses from a signal-generating block called a *clock*. The clock produces timed pulses. These pulses are sent to a counting circuit. The counting circuit output is a dc voltage. This dc voltage is developed from the counting information. Each unit of time is converted into a specific voltage. An exact voltage value, which represents some time frame, is sent to a decoder block.

The output of the decoder block is the dc voltage. This value is sent to an amplifier. The amplifier is required because of the low power level of the output of the decoder. The amplifier, in turn, operates an electromagnetic relay or a thyristor circuit. The control of the load occurs with the action of the relay or thyristor.

Relay Control

Electromechanical devices called relays have been utilized for many years as control devices. Each relay is rated in two ways. One way refers to the specific power requirements for the relay coil. The second refers to the contact arrangement and current rating. A typical power control relay is shown in Figure 20-13. Relays used for power control usually are designed to operate from an ac source. Typical operating voltages for the relay coil are 24 or 120 volts. The contact assemblies are rated in amperes. The actual rating is determined by the demands of the load. Contact assemblies may be arranged so that the circuit is turned on. These are called single-pole normally open contacts. The term single pole refers to the number of sets of contacts on the relay assembly. Double pole indicates two sets of contacts. A four-pole relay has four sets of contacts. The term "normally open" refers to the contact arrangement when the relay is de-energized, or at rest. These contacts are not connected in the relay "off" position. The term "normally closed" means that the contacts are touching when the relay is not energized.

Figure 20-13. Typical relay used for circuit control.

One example of a relay control system is illustrated in Figure 20-14. This is a low-voltage control system for lighting. A series of 24-volt ac relays are used to control 120-volt lighting circuits. It is possible to control any 120-volt load with this type of circuit as long as the relay contacts have sufficient current ratings.

The system uses a transformer to reduce the control from 120 to 24 volts. The advantage of this type of system is that wiring and switches are able to use components with lighter ratings than those used with the 120-volt source through a set of relay contacts. The coil of each relay is connected to a switch and the secondary winding of the transformer. Remote control of each circuit is possible with this system. All the 120-volt power requirements are connected to a central distribution point in the building. This central point also contains the relays. Wiring from the control switches to the relays uses a small-gage wire. Normal wiring from a central point to each load is relatively short in length. Control circuit wiring is often much longer. A great cost reduction occurs in a large installation by use of this remote-control wiring system.

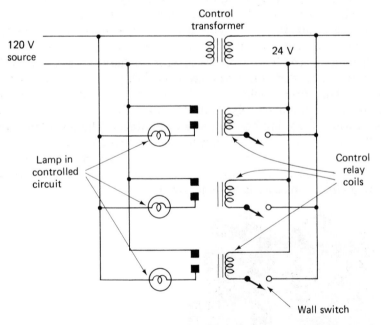

Figure 20-14. A low-voltage remote-control lighting circuit. Low-voltage relays control the power to the lamp circuits.

WAVE SHAPE CONTROL

The preceding section tells of a variety of methods of controlling the amplitude of an electrical wave. Either the voltage or the current, or both, are controlled with the systems. In each of the methods there is a reduction of power at the load. This power reduction is acceptable in some situations. In other cases a reduction of power produces a loss in the amount of work being performed. A machine that requires full power is not able to perform adequately under reduced power conditions. Therefore, another method is required in order to control this type of machine.

The power control that has become acceptable for high-energy machinery is waveform control. This is illustrated in Figure 20-15. Part (a) shows an in-phase voltage and current waveform and the resulting power curve. Both full and reduced amplitude voltage, current, and power waveforms are shown. It is easy to see that the reduced voltage produces less power than when the full voltage is applied to the circuit. Part (b) shows the applied voltage, current, and power under full-value conditions and when the time of the applied power is reduced.

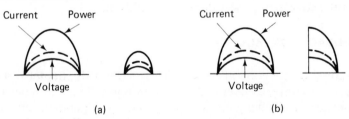

Figure 20-15. Waveforms for (a) voltage control and (b) waveform control. Full power is maintained with waveform control.

Reduction of applied power time does not produce less power. It does provide pulses of full power for shorter periods of time. This allows the load to operate at its full power rating. It is capable of performing all the required work when it is on. A system of this type is used in variable-speed electric drill motors. A typical schematic diagram for a motor speed control is shown in Figure 20-16. This circuit utilizes a triac and a control circuit in order to vary the on time of the load. The control circuit turns on the gate of the triac at some predetermined point in the ac cycle. The point of operation may be shifted by selection of the proper circuit components. The result is a

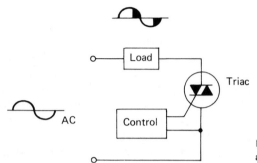

Figure 20-16. Triac control of an AC power circuit.

circuit that may be turned on for any portion of its cycle. This provides infinite control of the time that power is applied to the load.

The type of circuit described is used in drill motors because of the demand for a high torque when drilling. The variable-speed requirements cannot be met under reduced voltage and current conditions. The drill motor does not develop sufficient torque to do its work. The use of a thyristor waveform timing circuit meets the requirement for full torque. Speed of drilling (or cutting) is varied by the length of the duty cycle of the motor. This circuit is much more successful under high power demand conditions.

FEEDBACK CONTROL

The final power control system described in this chapter is a feedback type of control system. This system really does not have a start point because each element in the system is dependent on all the other elements. Look at the system in its forward (left to right) flow path as illustrated in Figure 20-17.

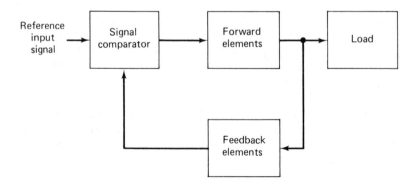

An input reference signal is established. This reference signal is often a voltage. It may be either an ac or a dc voltage value. The reference signal may even be a set of pulses from a pulse generator. The next block in the system compares two signals. One signal is from the input reference. The other signal is an error-correction signal. The output of this block establishes a voltage or signal that is used as an input signal to the control block. This signal may turn on a transistor or thyristor or it may be used to establish a specific operational voltage for the controller. The final block in the system is the load. This is the block in which the electrical power performs its work.

The significant block in this closed-loop system is the feedback block. This block has several names. Among these are feedback amplifier and error-signal amplifier. The purpose of this block is to sample a portion of the signal used to control the output. This signal is then returned to the signal comparator block. The input signal and the feedback signal are electronically added. The sum of these signals is used as the reference to set the control signal level. Should the feedback signal be a different value than the reference signal, the output of the comparator block modifies its command to the controller. Signals are continuously compared until the correct output signal is established.

This feedback control system may use mechanical, electrical, or electronic components. Some control systems use a combination of all three components. In any case, the systems are similar. Each operates in the same manner. A reference signal is established. The operational values of the load are used to monitor its operation. A portion of the output is fed back to a comparator block to maintain the correct operating conditions.

An example of a relatively simple feedback system is shown in Figure 20-18. It is the feedback system used to control heat in a building. Initially, a heat control is set at the desired temperature. A

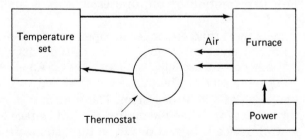

Figure 20-18. A feedback system used in the home for control of heat.

sensing device called a thermostat is used to control the duty cycle of a furnace. When the air temperature is below the desired setting, the thermostat contacts close. This turns on the furnace. The furnace produces heat. A portion of the heat is returned to the thermostat. When the temperature of the room (or building) reaches the level established on the thermostat, its contact points open. This turns off the furnace. The feedback path in this system uses the heated air. The heated air in the room is compared to the preset temperature position on the thermostat. When both conditions are equal, the furnace does not run.

This is an example of a simple feedback system. More complex systems are in use in homes, automobiles, and industry. Some examples of these include oven temperature settings, automobile "cruise" controls, and machine tool speed or positioning controls.

SUMMARY

One important electrical concept deals with the ability to control electric power. Control has to be done at the load end of a circuit in most cases. There are many different systems used to control electric power. These are grouped as amplitude control, wave shape control, and feedback control.

Amplitude control relates to the ability to vary the amplitude of the voltage or current in a circuit. This, in turn, varies the power in the circuit. Amplitude control is accomplished by varying the resistance in the circuit. Either a resistor or a resistive device like a transistor can be used for this purpose. Current control is accomplished by use of a variable inductance in the circuit. A movable core inductor does this very well.

Another way of controlling electric power is by use of a timer. Both mechanical and electric timers are used. The timing mechanism establishes an on-off sequence for the power to the load. A mechanical timer operates a disc, which, in turn, has a set of contact strips on its surface. The rotary motion of the disc establishes a timing sequence for the load.

Electrical timers use this principle to operate a switch. The sequence may be continual or it may be a one-time duty cycle.

Still another timer control uses electronic principles. An electronic clock is used to develop pulses. The pulses are modified to produce an operational voltage for a control system. The control system may be a transistor or a thyristor. When the desired control voltage level is reached, the control device is turned on. It, in turn, connects power to the load. A modification of this system uses the charge rate of a capacitor in a RC time circuit to accomplish the same results.

Electromechanical devices called relays are also used for power control. The relay has two circuits. One of these is its coil circuit. The coil forms an electromagnet when it is energized. The electromagnet moves an armature on which contact points are attached. The contact points form a part of the controlled circuit path with the load. Energizing the coil moves the contacts and completes the control circuit to the load.

Thyristors are used to turn on the control circuit of the load. The advantage of the thyristor is that it does not have any moving parts. It also can be turned on during a part of the ac cycle. This maintains full circuit power during the duty cycle. Pulses of power will reduce rotational speed while maintaining full operational torque in the load.

The final control circuit is a feedback system. A portion of the control signal is fed back to an input comparator circuit. The two signals are compared in order to establish a control signal value. When the feedback signal matches the reference signal, the load operates under correct conditions. Incorrect operational values are compared and adjusted in the comparator unit to maintain the proper operation of the load.

QUESTIONS

20-1. Amplitude control refers to the control of

    a. voltage levels

    b. current "on" time

    c. power "on" time

    d. resistance levels

20-2. Amplitude control is done by use of

    a. resistive networks

    b. rheostats

    c. variable resistors

    d. all of the above

20-3. Current control may be done by use of

    a. variable inductors

    b. motor generators

    c. manual starters

    d. all of the above

20-4. A device that controls the on-off cycle is called a

    a. rheostat

    b. variable resistor

    c. cycler

    d. timer

20-5. Electrothermal controls use the principle of

    a. mechanical movement

    b. electron flow

    c. heat

    d. rotary motion

20-6. Electromechanical controls use the principle of

    a. electron flow

    b. heat

    c. rotary motion

    d. thermal energy

20-7. Electronic controls use the principle of

    a. electron flow

    b. heat

    c. rotary motion

    d. thermal energy

20-8. Digital controls use the principle of

    a. electron flow

    b. heat

    c. rotary motion

    d. thermal energy

20-9. Wave shape control uses the principle of

    a. electron flow

    b. heat

    c. rotary motion

    d. thermal energy

20-10. Sampling of the output of a control and using it to make input adjustments is called

    a. return control

    b. feedback control

    c. output control

    d. none of the above

Index